AN APPLIED GUIDE TO PROCESS AND PLANT DESIGN

For Annemarie

AN APPLIED GUIDE TO PROCESS AND PLANT DESIGN

SEÁN MORAN

AMSTERDAM • BOSTON • HEIDELBERG • LONDON
NEW YORK • OXFORD • PARIS • SAN DIEGO
SAN FRANCISCO • SINGAPORE • SYDNEY • TOKYO
Butterworth-Heinemann is an imprint of Elsevier

ELSEVIER

ISBN: 978-0-12-800242-1

British Library Cataloguing-in-Publication Data
A catalogue record for this book is available from the British Library

Library of Congress Cataloging-in-Publication Data
A catalog record for this book is available from the Library of Congress
1007406946

For information on all Butterworth-Heinemann publications
visit our website at http://store.elsevier.com/

Typeset by MPS Limited, Chennai, India
www.adi-mps.com

Working together
to grow libraries in
developing countries

www.elsevier.com • www.bookaid.org

Publisher: Joe Hayton
Acquisition Editor: Fiona Geraghty
Editorial Project Manager: Cari Owen
Production Project Manager: Lisa Jones
Designer: Maria Ines Cruz

CONTENTS

Part 2 Professional Practice

Part 3 Low Level Design

Part 4 High Level Design

PREFACE

I am a highly experienced practical professional process engineer who has designed, commissioned, and undertaken troubleshooting of many process plants, and mentored and trained many other professional engineers in how to do these things. I am also a university professor who has taught process plant design to undergraduates and postgraduates for a number of years. This has required me to reflect upon what I know about the subject, how I know it, and how I can teach it to someone else. In this book I will assume the role of the experienced engineer, who takes lucky graduate chemical engineers by the hand in their first job or two and shows them what engineering is really about. Many are not so lucky as to have expert guidance. In doing so I will take the fairly informal tone I do when undertaking that task, and may on occasion express my frustration with the way the subject is taught in UK (and to the best of my knowledge worldwide) higher education. After all, it is not possible to describe how something might be improved without acknowledging that the present situation is less than perfect. I may also express the odd opinion and, as this is a distillation of experience rather than a scientific paper, I may not necessarily offer a reference to a peer-reviewed journal article in support of these opinions. However (despite the informal style of writing), the more controversial or provocative an opinion expressed, the more effort I have put into making sure that it is held by the majority of professional process plant designers. Toward this end, this book was reviewed by a panel of 40 professional engineers across sectors and worldwide. Many of the ideas which seem controversial in academic circles have been the subject of articles in The Chemical Engineer magazine, where they have been met with universally positive professional comment. The foundation of this book is practice, not theory. Throughout the text I will, however, offer quotations from others, links to books and even, on occasion, primary literature. These should not be misunderstood as the basis of my opinions. In the case of quotations, I am simply quoting people who agree with me. Suggestions for Further Reading are referenced to avoid my having to reproduce the content of these often weighty books, or reinvent the wheel.

ACKNOWLEDGMENTS

I would like to thank all of those who have helped me, especially my wife, Annemarie. Her patient assistance and sure hand with language have been essential in the preparation of this book. Heartfelt thanks also to Pat Cunningham for his painstaking proof-reading and advice on comma usage!

My fellow engineers have given generously of their time in helping me to make sure that what I have written represents consensus opinion, most notably: Gareth Brown, Chris Davis, Harvey Dearden, Mike Dee, Carlos Harrison, Myke King, Liangming Lee, Jim Madden, Brian Marshall, Keith Plumb, Ken Rollins, Michael Spreadbury, and Alun Rees. Thanks also to my colleagues George Chen and Giorgios Dimitrakis at the University of Nottingham for their comments and advice.

I am grateful to those who have kindly allowed me to reproduce images and material, including Tony Amato at Doosan Enpure, Sophie Brouillet at AMOT, John Evans at the Olympic Delivery Authority, Claudia Flavell-White at the IChemE, Kerry Harris at AUMA, Ernest Kochmann at Newson Gale, Malcolm Ledger at Lechler, Stuart Leigh and Ian Andrews at SLR Consulting, Edward Luckiewicz, Fiona Macrae at Crowcon, Glenn Miller at Grundfos, Ross Philips at the EEMUA, Jennifer Reeves at Elfab, Henry Sandler, Tosh Singh at Lutz-Jesco (GB) Ltd., Mike Wainwright at Ascendant and Kirsty Warren at WRAP.

The staff at Elsevier, especially Cari Owen, Lisa Jones and Fiona Geraghty, have been invaluable in making my experience as a neophyte to book publishing surprisingly painless.

PART 1

Practical Principles

Introduction

Process plant design is the pinnacle of chemical engineering design. Chemical engineering was developed based on the insight by Davis that all process industries used similar unit operations, which could be understood using sector independent analytical tools. Many "design tools" now commonly taught in academia incorporate assumptions that imply that all chemical engineering design is for the petrochemical sector, but chemical engineering has encompassed food and drink, inorganic chemical manufacture, and so on since its inception.

This book is about Process Plant design and, while examples may be drawn from my personal experience in the water and environmental sectors, it is intended to reflect consensus practice across the broad discipline. The IChemE's "Chemical Engineering Matters" discussion document identifies the energy, water, food and nutrition, health, and well-being sectors as the future of Chemical Engineering. We need to avoid confusion between chemical engineering and petrochemical engineering, a small and arguably diminishing subset of the discipline.

I have spoken and corresponded with hundreds of process engineers, most notably in the United Kingdom, Middle and Far East, in all industries to verify that the approaches suggested in this book do represent current consensus practice. It seems that professional practice has not changed very much over the last couple of decades, but that what is taught in universities has drifted further and further from professional practice during that period.

I am writing this book because I think that Chemical Engineering Education has lost its way, and become too theoretical and abstract to adequately serve its purpose, namely to provide the "academic formation of a Chartered Chemical Engineer" as the UK's IChemE puts it in its course accreditation guidelines.

The book is based on material I have delivered as part of the design courses I teach at the University of Nottingham, which is in turn based in my continuing professional engineering practice and professional training of my fellow engineers. It should be of use to undergraduate and postgraduate students, as well as early-career process plant designers and to university lecturers who wish to teach a more realistic version of plant design.

The book is in five parts. Firstly, I explain what process plant design is and how it is done in broad terms, then I give advice on professional practice in the most important aspects of process plant design, in general, and then at low and at high levels, and lastly I cover more advanced aspects of design.

It should be noted that this is a book about process plant design, rather than what is known nowadays in research-led universities as process design. There is no such thing as process design—processes happen in plants, and plants are the things which engineers design.

It has become clear in writing this book that we as a profession unhelpfully use the same words to mean different things. The meaning of the words and phrases Conceptual Design; HAZOP; Functional Design Specification; Design Philosophy; Design Basis; Process Intensification; Process Design; Optimization; Reproducibility; Repeatability; and Precision were particularly contested.

I have explained the sense in which I have used these words in the text at the point of first using them, and have included a glossary at the end. I am not claiming that my usage is the only correct one, but I have used them consistently in the book, and reflect to the best of my knowledge the most common meaning.

Seán Moran
2014

CHAPTER 1

Process Plant Design

INTRODUCTION

Whilst this may not be as obvious to today's students of the subject as it should be, chemical engineering is a kind of engineering, rather than a branch of chemistry.

Similarly, professional engineering design practice has next to nothing to do with the thing called process design in many university chemical engineering departments.

I will cover the reasons for this elsewhere, but first let's start by dispelling some confusion, by clearing up what engineering is (and is not), and what design is all about.

WHAT IS ENGINEERING?

I still feel glad to emphasize the duty, the defining characteristic of the pure scientist—probably to be found working in universities—who commit themselves absolutely to specialized goals, to seek the purest manifestation of any possible phenomenon that they are investigating, to create laboratories that are far more controlled than you would ever find in industry, and to ignore any constraints imposed by, as it were, realism.

Further down the scale, people who understand and want to exploit results of basic science have to do a great deal more work to adapt and select the results, and combine the results from different sources, to produce something that is applicable, useful, and profitable on an acceptable time scale.

C.A.R. Hoare

Engineers are those people "further down the scale" as Hoare the classicist and philosopher puts it, although I disagree that we "exploit the results of basic science." Our profession stands on other foundations, though you may have been taught something different in university.

In academia there is almost universal confusion between mathematics, applied mathematics, science, applied science, engineering science, and engineering. Allow me to unconfuse anyone so confused before we get started:

Mathematics is a branch of philosophy. It is a human construction, with no empirical foundation. It is made of ideas, and has nothing to do with reality. It is only "true" within its own conventions. There is no such thing in nature as a true circle, and even arithmetic (despite its great utility) is not factually based.

Applied mathematics uses mathematical tools to address some real problem. This is the way engineers use mathematics, but many engineers use English too. Engineering is no more applied mathematics than it is applied English.

An Applied Guide to Process and Plant Design

Science is the activity of trying to understand natural phenomena. The activity is rather less doctrinaire and rigid than philosophers of science would have us believe, and may well not follow what they call the scientific method, but it is about explaining and perhaps predicting natural phenomena.

Applied science is the application of scientific principles to natural phenomena to solve some real-world problem. Engineers might do this (though mostly they do not) but that doesn't make it engineering.

Engineering science is the application of scientific principles to the study of engineering artifacts. The classic example of this is thermodynamics, invented to explain the steam engine, which was developed without supporting science.

Science owes more to the steam engine than the steam engine owes to Science.

L.J. Henderson

This is the kind of science which engineers tend to apply. It is the product of the application of science to the things engineers work with, artificial constructions rather than nature.

Engineering is a completely different kind of thing from all preceding categories. It is the profession of imagining and bringing into being a completely new artifact which achieves a specified aim safely, cost-effectively, and robustly.

It may make use of mathematics and science, but so does medicine if we substitute the congruent "medical science" for "engineering science." If engineering was simply the application of these subjects, we could have a more-or-less common first and second year to medical and engineering courses, never mind the various engineering disciplines.

Now that we are clear about what engineering is, let us consider what design is.

WHAT IS DESIGN?

Rather than being some exotic province of polo-necked professionals, the ability to design is a natural human ability. Designers imagine an improvement on reality as it is, we think of a number of ways we might achieve the improvement, we select one of them, and we transmit our intention to those who are to realize our plan. The documents with which we transmit our intentions are, however, just a means to the ultimate end of design—the improvement on reality itself.

I will discuss in this book a rather specialized version of this ability, but we should not lose sight of the fact that design is in essence the same process, whether we are designing a process plant, a vacuum cleaner, or a wedding cake.

Designers take a real-world problem on which someone is willing to expend resources to resolve. They imagine solutions to that problem, choose one of those solutions based on some set of criteria, and provide a description of the solution to the craftsmen who will realize it. If they miss this last stage and if the design is not realized, they will never know whether it would have worked as they had hoped.

All designers need to consider the resource implications of their choices, the likelihood that their solution will be fit for the purpose for which it is intended, and whether it will be safe even if it not used exactly as intended.

If engineers bring a little more rigor to their decision making than cake designers, it is because an engineer's design choices can have life and death implications, and almost always involve very large financial commitments.

So how does engineering design differ from other kinds of design?

ENGINEERING DESIGN

Engineering problems are under-defined, there are many solutions, good, bad and indifferent. The art is to arrive at a good solution. This is a creative activity, involving imagination, intuition and deliberate choice.

Ove Arup

Like all designers, design engineers have to dream up possible ways to solve problems and choose between them. Engineers differ from, say, fashion designers in that they have a wider variety of tools to help them choose between options.

Like all designers, the engineer's possible solutions will include approaches to similar or analogous problems which they have seen to work. One of the reasons why beginners are inferior to experts is their lack of qualitative knowledge of the many ways in which their kind of problems can be solved, and more important still, those ways which have been tried and found wanting.

Engineers need to make sure they are answering the right question. For example, a UK missile program called "Blue Streak" was a classic engineering failure because the problem was not correctly stated. It was designed to be a long-range missile for nuclear warheads, but the missile had to be fuelled immediately before launch and it took thirty minutes to do this. Hence the missile was useless for the intended purpose, as it was not capable of sufficiently rapid deployment.

In "To Engineer is Human", Henry Petroski discusses the importance of avoiding failure in engineering design. Many of his examples of failure, however, were caused not by misspecification, but by designers who forgot that the models used in design are only approximations, applicable in a fairly narrow range of circumstances.

Billy Vaughn Koen goes still further toward the truth, when he points out that "all is heuristic." Even arithmetic is a heuristic. There are no absolute truths in mathematics, science, or engineering. There are only approximations, probabilities, and workable approaches. Engineers may just be a little clearer about these issues than mathematicians and scientists, because our solutions absolutely have to work.

PROJECT LIFE CYCLE

The niche or niches into which process plant design fits exist in the wider background of an engineering project life cycle. Engineers *conceive, design, implement,* and *operate* (CDIO) engineering solutions, as the CDIO Initiative points out.

The details of project life cycles vary between industries, but there is a common core. For example, here is the product life cycle for a pharmaceutical project:

1. Identify the problem—this stage is frequently overlooked because people think they know what the problem is. In reality, many solutions are for problems that do not exist (academic research often focuses on finding solutions without associated problems).
2. Define the problem in business, engineering, and scientific terms—often done poorly with the problem again being defined in terms of perceived solutions.
3. Generate options that provide potential solutions to the problem.
4. Review the options against agreed selection criteria and eliminate those options that clearly do not meet the selection criteria.
5. Generate the outline process design for the selected options.
6. In parallel:
   a. Commence development work at laboratory scale to provide more data to refine the business, engineering, and scientific basis of the options.
   b. Commence an engineering project to evaluate the possible locations, project time scale, and order of magnitude of cost.
   c. Develop the business case at the strategic level.
7. Based on the outcomes of step 6, reduce the number of options to those carried forward to the next level of detail.
8. In parallel:
   a. Continue the development work at the pilot plant scale.
   b. Based initially on the data from the laboratory scale, develop the design of the remaining options to allow a sanction capital cost estimate to be generated and a refined project time scale.
   c. Continue to develop the business case leading to a project sanction request at the appropriate corporate level.
9. Based on the outcomes of step 8, select the lead option to be designed and installed.
10. In parallel:
    a. Continue the development work at the pilot scale.
    b. Carry out the detailed design of the lead option. A "design freeze" will almost certainly need to occur before the development work is complete.
11. Construct the required infrastructure, buildings, etc. and install the required equipment.
12. Commission the equipment.

13. Commission the process and verify that the plant performs as designed and produces product of the required quality.
14. Commence routine production.
15. Improve process efficiency based on the data and experience gained during routine production.
16. Increase the plant capacity by making use of process improvements and optimization based on the data and experience gained.
17. Decommission the plant at the end of the product life cycle.

The pharma sector tends to run more stages in parallel than other sectors for reasons discussed later, but most of these stages exist in all sectors.

Where does design fit into this? Consultants might call stages 1−3 above plant design. Those with a background in design and build contracting, like me, usually think of design as being predominantly what those in operating companies call "grassroots design," broadly stages 3−10 above. Those who work for operating companies might call stages 15 and 16 plant design.

I suppose an argument can be made for all the above, but it should be noted that before step 14, very limited design information is available. The "design tools" popular in academia are used professionally only for stage 15/16 plant design, rather than stage 3−10 "grassroots" plant design for this reason.

I am going to use the term "process plant design" in the sense of stage 3−10 design throughout this book, even though stages 1−3 and stages 15 and 16 are certainly related fields of professional activity which I have been involved in. This is because this kind of design is closest to the meaning of the terms outside chemical engineering and there are several reasons for this.

Firstly, a piece of UK legislation called the Construction Design and Management (CDM) regulations require a person or company to declare themselves the designer of a plant, responsible for the safety implications of the design. This designer is almost always the entity responsible for stage 3−10 design.

Secondly, the process guarantee is almost always offered by the company responsible for stage 3−10 "grassroots" design.

Thirdly, this definition of design is that used by more or less everyone involved in design activity other than chemical engineers.

Fourthly, it's my book, and there are already lots of books on stage 15/16 "process design." This is, as far as I know, the only current book on "grassroots" process plant design.

PROCESS PLANT DESIGN

In researching this book, I looked at what had been published in texts intended to describe a process plant design methodology. There were promising sounding books with titles like "The art of…," "A strategy for…," "Systematic methods for…," "Design of simple and robust process plants," etc. I know that students and early stage designers lack an understanding of these things, as well as a gestalt of systems, but the overwhelming majority of these books failed to meet the promise of their titles.

Process plant design is an art, whose practitioners use science and mathematics, models and simulations, drawings and spreadsheets, but only to support their professional judgment. This judgment cannot be supplanted by these things, since people are smarter than computers (and probably always will be). Our imagination, mental imagery, intuition, analogies and metaphors, ability to negotiate and communicate with others, knowledge of custom and practice and of past disasters, personalities, and experience are what designers bring to the table.

If more people understood the total nature of design they would see the futility of attempts to replace skilled professional designers with technicians who punch numbers into computers. Any problem a computer can solve isn't really a problem at all—the nontrivial problems of real-world design lie elsewhere.

Engineering problems will almost certainly always be far quicker to solve by asking an engineer, rather than by programming a computer, even if we had the data (which we can never have on a plant which hasn't yet been built), a computer smarter than a person, and a program which codes real engineering knowledge, instead of a simplified mathematical model with next to no input from professional designers.

I wonder how the medical profession would feel if scientists and mathematicians suggested, without consulting medics, that they could produce an expert system which would exceed the competence of doctors?

This is a classic academic purist's mistake: The psychologist claims that sociology is just applied psychology; the biologist says that psychology is just applied biology, the chemist that biology is just chemistry with legs, the physicist that chemistry is just applied physics, the mathematician that physics is applied mathematics, and the philosopher that mathematics is applied philosophy. Emergent properties are irrelevant to the theorist, but in practical matters they may be everything.

Donella Meadows explains, in "Thinking in Systems," an intuitive system level view which is identical in many ways to the professional engineer's view. We share this view with the kindergarteners who also excel at a design exercise called the "Marshmallow Challenge" which I use in my teaching (see Further Reading). The roots of this system level view are natural human insights, which we may be educating out of our students.

Meadows explains that she makes great use of diagrams in her book because the systems she discusses, like drawings, happen all at once, and are connected in many directions simultaneously, whilst words can only come one at a time in linear logical order.

Process plant design is system level design, and drawings are its best expression—other than the plant itself—for the same reasons given by Meadows.

PROCESS PLANT VERSUS PROCESS DESIGN

Experimental scientists today, despite Einstein and Darwin, seem loath to abandon the search for an eternal changeless unhistorical reality of which pure mathematics could be the model.

Gordon Childe

An inward-looking school of "process design" as a form of applied mathematics has arisen in elite research institutions, whose practitioners collaborate only with their fellow researchers. They build upon each other's work, but their outputs are not used by, or indeed of use to, the profession.

Extending this philosophy to teaching programs, many universities have replaced essential professional knowledge with modules in which students learn to use researcher software so that, later in the course, they can carry out "process design" as these researchers do it.

Adherents of this school of thought argue that it is the job of industry to produce engineers, whilst academia's job is to provide an education in applied mathematics. They would argue that, not only should we follow institutions such as Tokyo University in not teaching our students to read engineering drawings, but they should not even spend much time learning about science.

The prevalence in academia of this approach based on modeling, simulation, and mathematical techniques such as network analysis seems to some extent to be an artifact of the way research is funded. Research which takes place in a PC rather than a laboratory (or worse yet a pilot plant) is relatively cheap to conduct, and thus to fund.

It is of course a valid function of engineering research departments to develop new design methodologies. Some aspects of these approaches may have niche applications in professional practice, but the overwhelming majority will not be taken up by the profession, as they do not help the professional to achieve their aims.

It is inevitable that researchers will consider their work important, even vital, but if it is not of use to the profession, it is research material of academic interest only. In my view, its most useful place (if any) within the UK model of chemical engineering education is only during the masters' year, when the Institution of Chemical Engineers (IChemE) requires students to receive greater exposure to research.

Variation and selection

Variation/creativity

When I examined myself and my methods of thought, I came to the conclusion that the gift of fantasy has meant more to me than my talent for absorbing positive knowledge.

Einstein

The engineering design process consists of the generation of candidate solutions, and of then selecting from these those most likely to be safe, cost-effective, and robust.

Coming up with the candidates is a creative process, involving the use of imagination, analogy from natural or artificial systems, knowledge of the state of the art, and so on. Selection of candidates is, however, often more of a grind.

Engineers will almost always make use of mathematical tools to help them with selection, though the academic study of engineering can overemphasize these tools, which are supposed to inform professional judgment rather than supplant it.

In "Engineering and the Mind's Eye," Ferguson points out a certain intellectual snobbery common in academia which values the mathematical over the verbal, and both of these over the visual, to the detriment of what we might call visual intelligence. Chemical engineering students may leave university with little or no drawing ability, or any development of their "visual intelligence."

It is left by many academics to their employers to teach new engineers this essential part of their skillset. This is bad enough, but perhaps the greatest problem with engineering education is slightly broader: the lack of opportunity to exercise creativity.

In a group design exercise previously referred to, known as the "Marshmallow Challenge," kindergarten children have been shown to outperform many engineering graduates in a test of practical imagination, visual reasoning, feel for materials, and group work.

I used to teach my students a risk management tool called Hazard and Operability Study (HAZOP; described later in this book) early in their course of study. All of them could master the formal methodology, but very few indeed could realistically imagine what might happen if a component failed (I have included at Appendix 2 a table to help with this).

Similarly, we used (as many institutions still do) to teach students how to use modeling and simulation programs such as Hysys in place of teaching process design. The vast majority of our students never learned to carry a mental model of a complex system in their heads, nor did they understand the limitations of the programs. They were Hysys operators, not engineers. This is not to say that modeling and simulation programs are worthless, but that they are being misused in academia, to the detriment of our students' understanding.

There is some evidence to suggest that this is a subset of a more general problem, in which the arguably too-ready availability of IT means that less is held in memory by younger people, and their visualization and mental modeling abilities are suffering as a result.

There are many formal systems intended to enhance creativity, but there is little evidence to suggest that they do more than ensure that a larger number of approaches are considered than might be if a less formal approach was followed. I find in my teaching that a greater degree of life experience is a better predictor of number of candidate solutions generated than the degree of adherence to a formal creativity enhancement methodology.

Selection/analysis

Modeling and simulation programs only address this part of the design process, and may allow thoughtless processing of options rather than making a genuine considered choice between well-understood options. Some say we are making technologists rather than engineers with this approach.

Tools which help the understanding of a system are good. Tools which allow bypassing of understanding to get to a decision (even if our understanding is "mere" intuition rather than science) are not. Our best students will still understand, but the generality will not and there are risky implications associated with this.

Simulation programs are not usually written by professional chemical engineers. They are usually written by numerate graduates with no plant design or operating experience, and they run necessarily simplified mathematical models on machines with a great deal of less processing power than people. Anyone who has seen the movie "The Terminator" knows what happens when machines build machines!

Professional engineers validate simulation programs against real plants before giving their outputs any credence. This is why the use of such programs by professional engineers is usually limited to modification of exiting plant rather than the whole-plant grassroots design addressed by this book.

ACADEMIC VERSUS PROFESSIONAL PRACTICE

In universities, students mainly practice alone or in small single-discipline groups with simplified and idealized examples. In order to turn the vague messiness of real engineering problems into a collection of unambiguous tasks, a great deal of data is given, the problem is very tightly framed, and in a "successful" example (judged by student satisfaction), it is very clear to students that they are faced with carrying out a number of the tasks which they were previously trained in.

Though these are very often supposedly group exercises, they are structured so that students can readily split them into a number of standalone tasks. When we examine students who have carried out such exercises, it becomes clear that none but the most able have any understanding of the overall picture—the majority are just grinding through the textbook method, or operating the program. Neither do they have any appreciation of the complexity of the substructures of a process plant. The method has produced neither system-level vision, nor an engineer's grasp of detailed considerations.

This may be useful as a way to learn design calculation techniques, or to illustrate basic principles, but this is not how the majority of engineering design is done. Pugh (see Further Reading) calls these academic approaches "Partial Design," which he contrasts with the professional's "Total Design."

Students mostly learn their basic science and math from professional researchers, with a very deep knowledge of a necessarily small area, and in the main a love of radical novelty. Engineering practice is far more holistic, and those who practice tend to dislike excessive novelty, which they see as inherently risky.

They also tend to dislike too much prediction of real-world outcomes based on scientific or mathematical models. They would rather see a full-scale working example of what is proposed: a model is not the thing itself. Professional engineers design things that absolutely have to work, and usually cost a great deal of money. Efficacy is a great deal more important than novelty.

This day-to-day activity of most practicing engineers is what Kuhn might call "normal design," where most designs are at best an incremental improvement on what has gone before, with a correspondingly small potential downside.

Practicing engineers operate in a highly constrained environment, and they understand that many of the problems inherent in a design scenario are simply too complex or vaguely defined to be analyzed rigorously within the resources available.

The design dimension sees engineering as the art of design. It values systems thinking much more than the analytical thinking that characterizes traditional science. Its practice is founded on holistic, contextual, and integrated visions of the world, rather than on partial visions. Typical values of this dimension include exploring alternatives and compromising. In this dimension, which resorts frequently to non-scientific forms of thinking, the key decisions are often based on incomplete knowledge and intuition, as well as on personal and collective experiences.

Figueredo

Pahl and Beitz's "Systematic Approach" (see Further Reading) splits the challenges of a design problem into these two components: uncertainty and complexity. According to them, if it is neither too uncertain nor too complex to be solved using standard design tools or methodologies, it is not a genuine problem at all, but it is merely a task.

Standard methods are of no use when there is insufficient data to apply them. There may well be some "tasks," such as checking the sizes of unit operations using heuristics, but these are mathematically trivial, and the product may well be ambiguous. There is no opportunity in professional life to start thinking that engineering is a branch of applied mathematics, or that a computer can solve engineering problems—computers only carry out tasks. Common sense is what is needed, and a feeling for ambiguity, qualitative knowledge, and multidimensional evaluation of options: in short, professional judgment.

The key to a successful design is to understand the problems just well enough to be able to predict that the desired outcome will be reliably attainable. Science and mathematics are certainly tools in the engineer's portfolio, but they are very often not the most important ones, and they are not its basis.

Capturing nonscientific information about how past designs of the type being attempted have performed, and the factors associated with success, is often at least as important. The information is highly situation-specific, and is essentially a codification of experience.

The information may be presented as a design manual, or as standards, codes of practice, or rules of thumb. Such documents allow the very complex situations common in real-world design to be appropriately simplified, making theoretically impossible parts of the design practically possible. The main purpose of such documents is to control the design process such as to constrain innovation within the envelope of what is known to be likely to work based on experience.

In researching this book, I found books by nonchemical engineers who clearly understand all that I have said so far, but for some reason, the link between professional practice and academic understanding seems to have been broken entirely in my discipline.

Process plants versus castles in the air

An architect friend tells me that there are two kinds of architects, those who know how big a brick is, and those who do not. Castles in the air are such stuff as dreams are made of, but real designs need to be built using things we can buy.

Process engineers do not often make things out of bricks, but we nevertheless make most of what we build from simple standardized subunits, such as lengths of pipe, and ex-stock valves, pumps, and so on. These items have types and subtypes which differ from each other in important ways. Choosing between them is no trivial matter. The essential but qualitative knowledge required to differentiate between these things is thought, by those who insist that only pure theory matters, to be beneath the dignity of universities to teach.

If you do not know the dimensions and characteristics of the standard subunits from which you build something, your designs will be impractical to build. They will be likely to be less cost-effective, less robust, and less safe than the output of a more practically minded designer.

The commercially available components almost always represent the lower limit of the professional engineer's resolution. We do not care about the atomic structure of the metal in our pump, but we care about those of its properties which affect the overall design.

What we design and what we do not

We don't tend to design things which we can buy, because having things specially made costs a lot more than buying stock items, and those who design things guarantee their efficacy. Engineers like to minimize both costs and risks.

Process engineers tend not to carry out design of mechanical, electrical, software, civil, or building works. They do, however, have to know about the constraints imposed on designers in these disciplines, and have a feel for the knock-on cost and other implications of their design choices.

Engineers mostly put standard components together in fairly standard ways. This doesn't sound very clever, until you consider that 32 chess pieces arranged on 64 squares following a set of simple rules can be positioned in 10^{50} legal ways. Engineers have to manage a similar level of complexity to that handled by chess players when designing a process plant. Standard parts and standard methodologies do not reduce the plant designer's profession to monkey work. They just make it humanly possible to reliably produce a working outcome.

In the academic setting, many students are taught engineering science in exercises intended to teach science in context. This is fine, but should not be mistaken for actually teaching engineering. Such academic exercises, which make engineering design look like applied science, have been stripped of their true complexity. Often there is only one "right answer" in such exercises. This is neither engineering nor design.

Standards and specifications

As practicing engineers we do not design, we specify. Specifications have the collective experiences over many years. They include successes and failures, and ultimately they stop us from killing people.

Anonymous Oil and Gas Process Engineer

Standards and specifications exist to keep design parameters in the range where the final plant is most likely to be safe and to work. They also serve to keep design documentation comprehensible to fellow engineers. A brilliant design which no one else understands is worthless in engineering.

There are a number of international standards organizations—ISO in Europe, DIN in Germany, ANSI, ASTM, and API in the United States, and so on ("British Standards" in the United Kingdom are now officially a subset of ISO).

Since I work in a UK University, I am going to refer to British Standards in this book where available, most notably those governing engineering drawings. The use of British Standards (or any other used consistently and clearly) for drawings reduces the likelihood of miscommunication between engineers via their most important channel of information exchange.

The availability of interchangeable standard parts makes much of engineering design simple in one way, but introduces an extra stage which can often be omitted in academic practice. After an approximate theoretical design, practitioners redesign in detail using standardized subcomponents. At this point, the very precise-looking academic design can turn out to be very precisely wrong.

We do not, for example, use 68.9 mm internal diameter pipe, we use 75 mm NB (nominal bore), because that is what is readily commercially available. NB and its US near-equivalent NPS (nominal pipe size) are themselves specifications, rather than sizes.

Design manuals

Companies frequently have in-house design manuals which are a formal way to share the company's experience of the processes it most often designs. These manuals are not in the public domain, because they contain a significant portion of the company's know-how. They are jealously guarded commercial secrets.

Some national and international standards are also essentially design manuals—for example, PD5500 used to be a British Standard and is now more of a design note or manual which embodies many years of experience of safe design of a safety-critical piece of equipment (namely the unfired pressure vessel), despite being superseded in 2002 by a European standard (BS EN 13445).

Rules of thumb

Rules of thumb are a type of heuristic, and are usually very simple calculations which capture knowledge of what tends to work. As Koen points out, there are only heuristics in engineering.

Rules of thumb do not necessarily replace more rigorous (but still heuristic) analysis when it comes to detailed design, but their condensation of knowledge gained from experience provides a quick route to the "probably workable" region of the design space, especially at the conceptual design stage.

Rules of thumb are only ever good in a limited range of circumstances. These limitations have to be known and adhered to if they are to be valid (though experts might knowingly break this rule on occasion). Rules of thumb encapsulate experience, and are therefore better than first principles design.

It should be noted that simulation and modeling are a kind of first principles design, and cannot be used to generate valid rules of thumb.

It should also be understood that first principles design, generally speaking, doesn't work, and that literature examples of successful first principles design are often no such thing when investigated.

Approximations

All is approximation in engineering. If you think you are precisely right, you are precisely wrong. Engineers who grew up in the era of the slide rule know that anything after the third significant figure is at best science, rather than engineering.

We need to know how precise and certain we have to be in our answers in order to know how rigorous to be in our calculation. Very often, in troubleshooting exercises, knowing the usual interrelationships between a few measurements is all that is needed to spot the most likely source of problems.

Coarse approximations will get us looking in the right general area for our answers, allowing design to proceed by greater and greater degrees of rigor as it homes in on the area of plausible design.

Professional judgment

Douglas gives a figure in "Conceptual Design of Chemical Processes" of 10^5-10^8 possible variations for a new process plant design, and he leaves out many of the important variables. This is a far lower number than the possible positions in chess, but it is similar to the number of patterns memorized by an experienced chess player.

An engineer's professional judgment allows them to semi-intuitively discern approaches to problems which might work in a similar way to the experienced chess player. They will summarily discard many blind alleys which a beginner would waste time exploring, and include options which beginners would be unlikely to think of. They will know which simple calculations will allow them to choose quickly between classes of solution.

Consequently, experts can quickly achieve outcomes which less experienced practitioners might never arrive at. This judgment takes many years of practice to develop, but its development may be started in an academic setting, and this book will attempt to assist in this task.

STATE OF THE ART AND BEST ENGINEERING PRACTICE

Six blind men who had never seen an elephant were asked to see what they could make of one which had wandered into their village.
   The first man touched its leg, and concluded it was like a pillar
   The second man touched its tail and thought it like a rope
   The third who touched its trunk thought it was like the branch of a tree
   The fourth felt its ear, and thought it was like a fan
   The fifth felt its side, and thought it like a huge wall
   The sixth who felt its tusk thought it like a pipe
   They began to argue about who was right about the elephant. Each insisted that he was right.
   A wise man passing by asked them, "What is the matter?" They explained that they each thought the elephant was quite different in form.
   The wise man explained that all of them were correct, which made them happy and allowed them to stop arguing about who was right.

The man in this story was clearly also wise enough not to point out to the villagers that, as they were blind, their understanding was rather partial, and none had grasped what a whole elephant was like to a sighted person, let alone an elephant keeper. No one thanks you for being that wise.

When I serve as an expert witness, the court requires me to differentiate between fact (perceptible directly to human senses or detectable with a suitable calibrated instrument), current consensus opinion (that held by the majority of suitable qualified professionals), and other opinion. These are held by the court to represent progressively weaker evidence. UK civil courts have a lower standard of proof than our professional one (they are content with a mere "more likely than not"), but their definition of sound professional opinion is a good one.

Koen essentially denies the existence of facts in engineering from a rather philosophical point of view, but he defines best engineering practice rather similarly, as current consensus professional opinion (essentially that part of a Venn diagram which would represent the overlap between the heuristics of all professionals).

I agree with these definitions, which means that best engineering practice is always changing, and no single engineer is an entirely reliable source of best engineering practice.

My experience as an expert witness has, however, taught me that we have, as professionals, an idea of the gap between our personal practice and common practice. We know whether each of our heuristics is on the fringes or at the heart of consensus/best practice. Even those maverick professionals who hold fringe ideas which they think better than best practice know that these opinions are not commonly held.

This means that only current practitioners who regularly engage professionally with other practitioners are in touch with the moving target of best practice. It also means that those closest to the heart of a discipline tend to have the greatest concordance between personal heuristics and consensus heuristics. This does not prevent people far from the heart of the discipline from holding fringe opinions strongly; it just keeps them from understanding that they are fringe opinions.

THE USE AND ABUSE OF COMPUTERS

Back in 1999 the IChemE's Computer Aided Process Engineering (CAPE) working group produced Good Practice Guidelines for the Use of Computers by Chemical Engineers. These have never been superseded, but they seem to have been largely forgotten in the interim.

The guidelines emphasize the legal and moral responsibility of the professional engineer to ensure the quality and plausibility of the inputs to and outputs from the system, to understand fully the applicability, limitations, and embedded assumptions in any software used. It also emphasizes the importance of being properly trained to use the software, and only using fully documented software validated for the particular application.

They emphasize the primary importance of understanding the problem one is trying to solve, working within proper engineering limits, of taking into consideration not just dynamic but transient conditions and, most of all, of applying sensitivity analysis to the results produced, especially for those areas identified as the most important to a successful outcome. The severity of the consequences of being wrong should also be taken into consideration.

They warn users to beware of the assumptions implicit in software, and to know where default values exist, and to be able to back-trace any data used to a validated source. This allows the documented accuracy of the data in the range used to be known, as well as whether it is valid in that range.

They recommend contacting more senior engineers and/or the suppliers of the software in the event of any uncertainty at all about the correctness, accuracy, or fitness for purpose of any software or outputs.

In several places, they specifically recommend suspicion about the outputs of computer programs, and an assumption of guilt until proven innocent.

They say again and again, in different ways, that software cannot be a substitute for engineering judgment, and its use without understanding is a dangerous abandonment of professional responsibility.

However, I have seen many times, in both academic circles and in recent graduates, a failure to understand this simple truth. Computers may be a little faster than they were in 1999, but they are no closer to being people. All the potential problems of software are still there, and the awareness of their limitations has decreased. In my view, this is a disaster waiting to happen.

FURTHER READING

CDIO Initiative: See <www.cdio.org>.

Ferguson, E., 1994. Engineering and the Mind's Eye. MIT Press, Cambridge, MA.

IChemE CAPE Working Group, 1999. The Use of Computers by Chemical Engineers: Guidelines for Practicing Engineers, Engineering Management, Software Developers and Teachers of Chemical Engineering in the Use of Computer Software in the Design of Process Plant. Institution of Chemical Engineers, London.

Koen, B.V., 2003. Discussion of the Method. Oxford University Press, New York, NY.

Meadows, D.H., 2009. Thinking in Systems. Routledge, New York, NY.

Pahl, G., Beitz, W., 2006. Engineering Design: A Systematic Approach. Springer, London.

Petroski, H., 1992. To Engineer Is Human: The Role of Failure in Successful Design. Vintage, New York, NY.

Wujek, T., Undated. The Marshmallow Challenge. <http://marshmallowchallenge.com>.

Vincenti, W.G., 1990. What Engineers Know and How They Know It: Analytical Studies from Aeronautical History. Johns Hopkins University Press, Baltimore, MD.

CHAPTER 2

Stages of Process Plant Design

GENERAL

Process plant design (and indeed almost all design) proceeds by stages which seem not so much to be conventional as having evolved to fit a niche. The commercial nature of the process means that minimum resources are expended to get a project to the next approval point. This results in design being broken into stages leading to three approval points, namely, feasibility, purchase, and construction.

This is why Pahl and Beitz's systematized version of the engineering design process resembles that which applies to all engineering disciplines (including chemical engineering's process plant design) as practiced by professionals. It may very well also apply to fashion design. Design is design. Is design.

Note that in recommending Pahl and Beitz's approach I am not seeking to enter the academic debate on how the design process *ought* to be done. Having read many books on engineering design across many disciplines, I found Pahl and Beitz's description to be one of the closest to how design *is* done. That is the subject of this book.

The basically invariant demands of the process are the reason why everyone who designs something professionally does it basically the same way, even though chemical engineers are often nowadays explicitly taught a radically different approach in university (if they are taught any approach at all).

CONCEPTUAL DESIGN

Conceptual design of process plants is sometimes carried out in an ultimate client company, more frequently in a contracting organization, and most commonly in an engineering consultancy.

In this first stage of design, we need to understand and ideally quantify the constraints under which we will be operating, the sufficiency and quality of design data available, and produce a number of rough designs based on the most plausibly successful approaches.

I am told that, in the oil and gas industry, the conceptual stage starts from a package of information known as Basic Engineering Design Data (BEDD), which is often confused with (Process) Basis of Design. BEDD includes, typically, information to start the concept design such as:
- General plant description
- Codes and standards

- Location
- Geotechnical data
- Meteorological data
- Seismic design conditions
- Oceanographic design conditions
- Environmental specifications
- Raw material and products specifications
- Utilities
- Flares
- Health Safety Environment (HSE) requirements

Other industries have alternative formats, but initial information packages ideally cover many of the same areas.

Practicing engineers tend to be conservative, and will only consider a novel process if it offers great advantages over well-proven approaches, or if there are no proven approaches. Reviews of the scientific literature are very rarely part of the design process. Practicing engineers very rarely have the free access to scientific papers which academics enjoy, and are highly unlikely in any case to be able to convince their colleagues to accept a proposal based on a design which has not been tried at full scale several times, preferably in a very similar application to the one under consideration.

The conceptual stage will identify a number of design cases, describing the outer limits of the plant's foreseeable operating conditions. Even at this initial stage, designs will consider the full expected operating range, or design envelope.

The documents identified in Chapter 3 are produced for the two or three options most likely to meet the client's requirements (usually economy and robustness). This will almost always be done using rules of thumb, since detailed design of a range of options (the majority of which will be discarded) is uneconomic.

This outline design can be used to generate electrical and civil engineering designs and prices. These are important, since designs may be optimal in terms of pure "process design" issues like yield or energy recovery, but too expensive when the demands of other disciplines are considered.

At the end of the process, it should be possible to decide rationally which of the design options is the best candidate to take forward to the next stage. Very rarely, it will be decided that pilot plant work is required, and economically justifiable, but this is very much the exception; design normally proceeds to the next stage without any trial work.

There are academic arguments for including formal process integration studies at this stage, though this is incredibly rare in practice. The key factor in conceptual studies is usually to get an understanding of the economic and technical feasibility of a number of options as quickly and cheaply as possible. As many as 98% of conceptual designs do not get built, so you don't want to spend a fortune investigating them.

Client companies have advantages over contractors in carrying out conceptual designs, as they may have a lot of operating data unavailable to contractors, however, they do usually lack real design experience.

Contractors are in the opposite situation, while the majority of staff employed by many consultancies tend to have neither hands-on design experience nor operational knowledge.

In an ideal world, therefore, client companies would collaborate with contractors to carry out conceptual design. In the real world, this cooperation/information sharing is less than optimal.

"CONCEPTUAL DESIGN OF CHEMICAL PROCESSES"

Douglas wrote a book of this name which essentially attempts to design chemical processes (whatever they are), rather than process plants.

He understands that the expert designer proceeds by intuition and analogy, aided by "back of the envelope" calculations, but sees the need for a method which helps academics and beginners to cope with all the extra calculations they have to do while they are waiting to become experts (who know which calculations to do).

The arguments underlying the academic approach which has since been built on Douglas's approach are helpfully set out in explicit detail. There is an assumption that the purpose of conceptual design is to decide on process chemistry and parameters such as reaction yield. Choices between technologies (the usual aim of conceptual design exercises) are not considered. Pumps are assumed to be a negligible proportion of the capital (capex) and running (opex) cost, and heat exchangers are assumed to be a major proportion of capex and opex.

It is implicit in the chain of assumptions used to create the simplified design methodology that a particular sort of process is being designed. Like all design heuristics, the methodology has a limited range of applicability. While it mentions other industries, it is based throughout upon examples taken from the petrochemical industry, and it is clear that the assumptions it makes are most suited to that industry.

Having declined to consider many items which are of great importance in other industries, Douglas finds time for pinch analysis, which was quite new when the book was written. Perhaps this really was a worthwhile exercise for the novice process designer in the petrochemical industries of the 1980s, but there are many process plant designs in 2014 which do not have a single heat exchanger.

In the majority of industries, process chemistry is a job for chemists, and from the plant designer's point of view is in any case usually limited to choosing between a number of existing commercially available process technologies.

Douglas offers a plausible approach to the limited problem he sets out to solve, few of whose assumptions I can argue with in the context of his chosen example. He attempts to offer the beginner a way to choose between potential process chemistries and to specify the performance of certain unit operations in a rather old-fashioned area of chemical engineering.

However, the problem which this methodology seeks to address is not one I have ever been asked to find a solution for. When I am asked to offer a conceptual design, I am being asked to address different questions, on plants with a different balance of cost of plant components. Petrochemical plants of the sort used as the example in this book do not really get built in the developed world any more.

The approach does, however, hang together coherently, in a way more recent developments based on it do not. A good amount of effort goes into constructing as rigorous a costing as is possible at the early design stage (ignoring the issue of the items which are left out).

In essence this book seems, in my view, to reflect the slight wrong turns and over-simplifications which, followed by successive oversimplifications and misunderstandings, led to the utterly unrealistic approaches common in academia nowadays.

The attraction of the approach to academics is presumably that it is intended to allow people who have never designed a process plant (like the majority of academics) to use the skills they do have to approximate an early stage design process.

The approach is a tool to allow a very inexperienced designer who does not have access to expert designers to simplify the design of a certain sort of petrochemical plant to the point where they can mathematically analyze the desirability of a small number of parameters such as degrees of reactor yield and energy recovery.

This approach is just another design heuristic, and like all heuristics it has a limited range of applicability. If the very specific assumptions it makes to achieve this aim are not met, its use is invalid.

Even if they are met, process integration at conceptual design stage is both uneconomic and unwise, for reasons I discuss later in the book.

Modeling as "conceptual design"

Much work thought of as conceptual design, especially in the case of modifications to existing plant, takes place in large consultancies and operating companies. Because plant operating companies and consultants lack real whole-plant design experience, this may feature some elements of the academic approaches derived from Douglas's approach discussed in the last section, combining the application of network and pinch analysis to the output of modeling and simulation programs.

The scope of such studies is usually for a small number of unit operations rather than a whole plant, and it is by talking solely to those carrying out such studies (and each other) that academics get the impression that their techniques are used in industry. Academics tend to stay in touch with the kind of students who do well in exams and who are preferred by such operating companies.

However, this work differs from the design techniques of professional process plant designers (as I define them in Chapter 1) in a number of ways, most notably the following:

The "conceptually designed" plant will not be built by those carrying out the conceptual design exercise. The contractors who will build it may use this "design" as the basis of their design process, but since it is they who will be offering the process guarantee, they will redesign from the ground up. They will, however, probably not point this out to the client, who will consequently retain the impression that they designed the plant.

Genuine information from pilot studies on the actual working plant allows the model used to be fed with a specific and realistic design envelope, and for its outputs to be validated on the real plant. This is very different from using the modeling program in an unvalidated state.

Thus, such conceptual design studies, while potentially very valuable as a debottlenecking or optimization exercise, are not a true process plant design exercise. The modeling, simulation, and network analysis beloved of research academics are of their greatest utility in this area, due to the availability of the large quantities of specific data which such approaches require to yield meaningful results.

However, I would still suggest to those carrying out such exercises that they should be willing to listen to suggestions from contracting companies if they would like to arrive at a safe, robust, and cost-effective solution. There is no substitute for experience.

All should understand that professional judgment is still superior to the output of computer programs, however much effort went into the pilot trials and modeling exercises, but human nature leads people to wish to benefit from costs which have been irreversibly incurred. If you have spent a great deal of resources on these studies, it may be hard to accept that you could have simply got a contractor to design the plant without spending the resources. The sunk cost fallacy needs to be avoided.

FRONT END ENGINEERING DESIGN (FEED)/BASIC DESIGN

If the design gets past the conceptual stage, a more detailed design will be produced, most commonly in a contracting organization.

Competent designers will not use the static, steady state model used in educational establishments, but will devise a number of representative scenarios which encompass the range of combined process conditions which define the outer limits of the design envelope. All process conditions in real plants are dynamic; they do not operate at "steady state." They may approximate the steady state during normal operation if the control system is good enough, but they must be designed to cope with all reasonably foreseeable scenarios.

The process engineer would normally commence by setting up (most commonly in Excel) a process mass and energy balance model linking together all unit operations, and an associated Process Flow Diagram (PFD) for each of the identified scenarios. It is usually possible to set up the spreadsheets so that the various scenarios are produced by small modifications to the base case.

In certain industries (especially the oil and gas sector) a modeling system like Pro II might be used for this stage, but across all sectors this would be the exception rather than the rule, because business licenses for such programs are very expensive, and the available unit operations tend to be tailored to a specific industry sector.

A more accurate version of the deliverables from the previous stage would be produced, based on this more detailed design/model and, wherever possible, bespoke design items would be substituted with their closest commercially available alternatives, and the design modified to suit.

Process designers would normally avoid designing unit operations from scratch, preferring to subcontract out such design to specialists who have the know-how to supply equipment embodying the specialist's repeated experience with that unit operation.

Drawings at this stage should show the actual items proposed, as supplied by chosen specialist suppliers and subcontractors. Even such seemingly trivial items as the pipework and flanges selected should be shown on the drawings, as they are supplied by a particular manufacturer, and pricing should be based on firm quotes from named suppliers.

The drawings should form the basis of discussions with, at a minimum, civil and electrical engineering designers, and a firm pricing for civil, electrical, and software costs should be obtained.

All drawings and calculations produced would be checked and signed off at this stage by a more experienced—ideally chartered/licensed professional—chemical engineer.

Once that is done, a design review or reviews can be carried out, considering layout, value engineering, safety, and robustness issues. Where necessary, modifications to the process design to safely give overall best value should be made.

DETAILED DESIGN

Or "Design for Construction." This virtually always takes place in a contracting organization. The detailed design will be sent to the construction team, who may wish to review the design once more with a view to modifying it to reflect their experience in construction and commissioning.

Many additional detailed subdrawings are now generated to allow detailed control of the construction of the plant. The process engineer would normally not have much to do with production of such mechanical installation drawings, other than participation in any design reviews or Hazard and Operability Studies (HAZOPs) which are carried out.

The junior process engineer will, however, have much to do with producing documents such as the datasheets, valve, drive, and other equipment schedules, installation and commissioning schedules, and project program.

This painstaking work is required to allow the procurement of the items which are described in them by nonengineers. It is not so much design work as contract documentation.

SITE REDESIGN

This does not usually feature much in textbooks on the subject, but it is not uncommon for designs to pass through all the previous stages of scrutiny and still be missing many items required for commissioning or subsequent operation.

It is a lot cheaper to move a line on a drawing than it is to reroute a process line on site. If communication from site to designers is managed very poorly in a company, expensive site modifications may be required on many projects before the problem comes to the attention of management.

Commissioning and site engineers are rarely involved in the design process (though they should be involved in the HAZOP) but often find these omissions when they review the design they have been given, or worse still, find the error on the plant after it was built. This is an expensive stage at which to modify a design, but in the absence of perfect communication from site to design office, it will continue to be needed.

On behalf of commissioning engineers everywhere, therefore, I would like to encourage designers to make sure that the following items are included in designs at the earliest stage:

- Tank drains which will empty a tank under gravity in less than one hour, and somewhere for the drained content to go to safely,
- Tank vents or vent/vac valves (with protection where required) to allow air to enter and exit a tank safely,
- Suitable sample points in the line after each unit operation (there may be far more to this than a tee with a manual valve on it),
- Service systems adequately sized for commissioning, maintenance, and turnaround conditions,
- Connection points for the temporary equipment required to bring the plant to steady state operation from cold need to be provided, especially where process integration has been undertaken,
- Water and air pollution control measures required under commissioning conditions,
- Access and lifting equipment required under commissioning, maintenance, and turnaround conditions.

Designers should in my opinion also avoid the following features, which are likely to be removed by commissioning engineers or operators:

- Permanent pump suction strainers or filters, especially fine mesh filters.

There is a lot more on this in Chapter 17.

POSTHANDOVER REDESIGN

The commissioning engineer may have tweaked or even redesigned the plant to make it easier for it to pass the performance trials which are used to judge the success of the design, but the nature of the design process means that while the unit operations have been tuned to work together, they actually have different maximum capacities.

When the spare capacity in the system is analyzed, it is usually the case that the output of the entire process is limited by the capacity of the unit operation with the smallest capacity. This is a restriction of capacity or "bottleneck." Uprating this rate-limiting step (or a number of them) can lead to an economical increase in plant capacity.

Similarly, it might be that services which were slightly overdesigned to ensure that the plant would work under all foreseeable circumstances, and optimized for lowest capital cost rather than lowest running cost, can be integrated with each other in such a way as to minimize cost per unit of product. This is important, because prices for a plant's product tend to fall over its operating life, as competing plants which are built in low-cost economies or based on newer, better processes bring prices down.

It should be noted that those carrying out this kind of operation have at their disposal a lot more data than whole-process designers. Clever mathematical tools were developed back in the 1970s to facilitate the energy integration process, and their conceptual approach has since been applied to mass flows, including those of water and hydrogen.

These were devised to be optimization tools, rather than process design tools. As I discuss elsewhere in this book, there are a number of inevitable drawbacks to their misuse as design tools, which are more serious the earlier they are used in the design process.

This book does not have much to say about post-handover design activity, as it does not meet a number of my criteria to qualify as process plant design. Most notably, it does not involve designing a whole plant; there is a great deal of site-specific detailed information available to validate computer models; and small improvements in performance are within the resolution of the design process.

This activity is, to my mind, aftermarket tuning of a plant which has been designed by others, rather than designing one from scratch, the subject of this book.

UNSTAGED DESIGN

Sometimes it is more important to get a design completed quickly than to spend time optimizing cost, safety, and robustness. These things still matter, and have to be addressed to at least a minimum standard, but there can be commercial drivers which mean that getting the product of the process on the market as soon as possible is the most important factor.

The premier example of this would be design of systems producing a product which is subject to patent, especially pharmaceuticals. Once patent protection is removed, generic substitutes manufactured in low-cost/wage economies will usually rapidly out-compete the patented version.

Pretty much all the money which will be made on the product will come in between the day it hits the shelves and the day the patent expires. Thus, getting the product to market quickly is more important than plant design optimization, to the dismay of my colleagues who work in pharma.

PRODUCT ENGINEERING

European chemical engineering bodies such as FEANI are encouraging the promotion of "product engineering" into chemical engineering curricula. It should, however, be noted that "product engineering" is not the same thing as "product design," a term used in academia for something which is often not any kind of engineering.

Product engineering is the name given to shortening the classic approach described in this chapter and running it alongside a product design program. The deliverables of product engineering are identical to those listed earlier in this chapter. Price, practicality, and safety issues need to be as much at the forefront of the process as they are with the classic approach.

Process Plant Designers design coordinated assemblies of machines (gubbins) to make chemicals (stuff). Our plants make stuff, rather than gubbins, and our plants are made of gubbins rather than stuff. Mechanical engineers design gubbins, and chemists work with stuff.

Academic "product design" is often just rebranded chemistry content, with little consideration of plant or production cost, safety, or robustness. I have even seen it used to describe an exercise in which students were asked to research and design a university teaching module.

The design methodology of professional product design is just like that of the professional plant designer. Pugh's "Total Design" is about product design, but is far closer to a description of how process plant design is actually undertaken than almost all books by "chemical engineers."

FAST-TRACKING

Mixing the natural stages of design in order to accelerate a program is a reasonably common approach, which may be used in professional practice (though it has a downside). It is telling that contractors who are asked to make a program move faster (or "crash the critical path") are usually given acceleration costs to compensate them for the inherent inefficiency of the fast-tracked process.

As well as costing more for the reasons given below, fast-tracking is commonly held to always increase speed at the cost of quality, whether that be design optimization quality and/or quality of design documentation.

The standard approach practically minimizes the amount of abortive work undertaken, since each stage proceeds on the basis of an established design envelope and approach. Each stage refines the output of the preceding stage, requires more effort, and comes to a more rigorous set of conclusions than the preceding stage.

When stages are mixed, the more rigorous steps are carried out earlier in the program when the design has a larger number of variables. We may well therefore need to have a larger team of more experienced people working on the project. Even with the improved feel for engineering given by using more experienced engineers, it is much more likely in fast-tracking that a design developed to quite a late stage will need to be binned, and the process restarted from the beginning of the blind alley which the design went down. Of course, if you are in a sector where you are making an extremely profitable product, this will be less of a concern than if you are in a less lucrative field.

Generally, costs for product by sector may be ranked as follows:

Biopharmaceuticals
Pharmaceuticals/nanotech products and the like
Fine chemicals
Oil and gas
Bulk chemicals
Water and environmental

Oil and gas have an anomalous position in this hierarchy (and are highly profitable) because they produce huge volumes of a product whose price is effectively set largely by an international cartel.

Conceptual/FEED fast-tracking

In a contracting company or design house, conceptual design can be very quick indeed. Senior process engineers know from experience which is likely to be the best approach to an engineering problem. A few man-hours may be all that they require to rough something up.

However, this means that a client who wishes to skip the stage where they or a consultant do the initial designs is ceding control of the design to a contractor. This is not without a downside—senior process engineers in contracting organizations will almost certainly come up with a design which will be safe, cost−effective, and robust, but a design based entirely on generic experience is unlikely to be the most innovative practical approach.

FEED/detailed design fast-tracking

If a project is definitely going to go ahead, and the contractor has been appointed, FEED and detailed design can be combined, and the operating company/ultimate client can take part in the design process. This was quite a common approach using the Institution of Chemical Engineers (IChemE) "Green Book" contract conditions back at the start of my career.

This approach can produce a very good quality plant, which is unusually fit for purpose, but the FEED study is used as the basis of competitive tendering. The removal of a discrete FEED study, along with that of the usually adversarial approach between client and contractor in process plant procurement, does seem to inflate costs somewhat.

Design/procurement fast-tracking

Time is generally lost from the originally planned design program at each stage of design such that the later stages—which produce the greatest number of documents, employ the most resources, and are most crucial to get right—often proceed, unhelpfully, under the greatest time pressure.

There are barriers to communication between the various stages of design which need to be well managed if conceptual design is to lead naturally to detailed design and from there to design for construction. If the communication process is managed poorly, unneeded redesign may be carried out by "detailed" or "for construction" design teams who do not understand the assumptions and philosophy underlying the previous design stage. Alternatively designs may have to be extensively modified on site during construction and commissioning stages, usually at the expense of the contractor.

Design/procurement fast-tracking is quite popular as a response to time lost in earlier stages for projects where the end date cannot move. As soon as an item's design has been fixed, the procurement process is started, especially with long lead time items like large compressors.

The nature of design being what it is, this can result in variations to specifications for equipment design after procurement has started, and this usually carries a cost penalty, as does the high peak manpower loading, duplication of designs, and backing-out of design blind alleys inherent in this approach.

The fast-track to bad design

The "chemical process design" approach popular in academia attempts to mix all design stages prior to construction, and often substitutes modeling for construction (a mistake I comment on elsewhere in the book). It is frequently argued by those who teach this that it is how design *ought* to be done, as sustainability concerns mean that we need to approach thermodynamic limits to energy recovery.

The first problem I have with this is that sustainability is a highly politicized term. A Greenpeace member, a trade union activist, and a Chartered Chemical Engineer might use the term to mean three completely different things. The IChemE have helpfully written guidelines on what it means to chemical engineers, in the form of sustainability metrics. We engineers like a metric, so that we can analyze a problem and its possible solutions at least semi-quantitatively.

The IChemE interpretation does not support shooting for theoretical perfection in a small number of aspects of process design. Chemical engineers are concerned about the environment, but we know that the curves of process yield, energy recovery, safety, and environmental protection against cost are exponential. Both perfect processes and complete safety are infinitely costly.

My second problem is that it is not just that you cannot optimize a process before you design it in detail: you cannot really optimize a process before you have built it. At the conceptual design stage the design subcomponents are not items you can actually buy economically. There are a number of reasons for this which are as follows:

Firstly, sizes of available process equipment are not a continuous variable. There are minimum and maximum available sizes, and size increases in discrete steps. This is true of the very simplest items such as lengths of pipe. You can have special items made to order, but they are far more expensive than stock items, take far longer to deliver, and are more likely to have unforeseen problems.

Secondly, no plant is ever built exactly as specified. Sometimes this is due to errors in construction or poor QA in materials or construction but, more significantly, many plants need site-level redesign due to design errors, unforeseen consequences, or a late change in client specification.

Thirdly, this approach takes no consideration of the interactions with other disciplines (most notably civil and electrical engineering) of process design choices. These can be very significant, far more significant than the cost implications of theoretically tweaking a virtual plant for small increases in energy efficiency or yield.

Fourthly, it is commonplace in this approach to take no real consideration at all of the impact of process choices on either capital or running costs of the plant. If you aren't costing, you aren't engineering, but those who practice "chemical process design" apparently consider cost only by comparing the cost of feedstock with that of product. If product can be sold for more than feedstock can be bought, the process is deemed to be economic. Thus the marginal cost of yet another heat exchanger is set to zero, such that it is always viable to go one step closer to theoretical perfection.

I have been told that the application of aspects of this approach to real plant design has led to processes whose safety factors are so tight that it is thought unwise by experienced engineers to attempt debottlenecking, though management still expects it.

This is arguably what the approach really does: any savings which might be found will come out of safety factors and plant operability. It all seems very reminiscent to me of the erosion of safety factors which presage disaster in many of Henry Petroski's examples. It all seems terribly clever until your bridge falls down, or your plant is replaced by a crater.

All of this said, there are circumstances where simulation can be used to partially replace design: if you have a great deal of operating data for exactly the plant you are "designing," you can tune and validate the model, such that you are no longer designing from first principles and generic data, but have an empirically verified simulation of the actual plant.

This is the case if you are working for an operating company or a contractor/operator with access to such data, and the spare time necessary to tweak the simulation. However, you may not necessarily understand why the simulation behaves as it does, even in the case where you seem to have made it behave exactly like the real plant.

If you do not understand your model, it is worse than useless.

FURTHER READING

Azapagic, A., 2002. Sustainable Development Progress Metrics Recommended for Use in the Process Industries. Institution of Chemical Engineers, London.

Pahl, G., Beitz, W., 2006. Engineering Design: A Systematic Approach. Springer, London.

Petroski, H., 2012. To Forgive Design. Harvard University Press, Cambridge, MA.

Pugh, S., 1990. Total Design. Prentice Hall, Upper Saddle River, NJ.

CHAPTER 3

Process Plant Design Deliverables

OVERVIEW

"Deliverables" may not be the most elegant word, but it is freighted with meaning. It comes from project management, and it reminds us that the immediate purpose of the design process is to deliver to a client a set of documents which they can use to build a plant, or more usually approve formally, so that the designer's company can build it.

Less frequently, the plant itself is described as a deliverable, but we will restrict ourselves in this section to the drawings and other documents which are commonly used to transmit design and construction intent from the designer to the construction team.

The following sections list these documents in the rough order in which they are first produced, although revision of such documents may be ongoing throughout the project.

These are only the most important and commonplace deliverables; I am deliberately omitting many sector-specific deliverables.

DESIGN BASIS AND PHILOSOPHIES

The output from the conceptual design stage may sometimes be restricted to guidance on the approach which should be followed in subsequent design stages: a *design basis* or *design philosophy*.

These terms are sometimes taken to be the same thing, but I will differentiate between them as follows:

In professional practice, a design basis will usually be a succinct (no more than a couple of sides of A4) written document which might define the broad limits of the Front End Engineering Design (FEED) study, including such things as operating and environmental conditions, feedstock and product qualities, and the acceptable range of technologies.

Design Philosophies by contrast may run to 40 pages, including overpressure protection philosophies, vent, flare, and blowdown philosophies, isolation philosophies, etc.

Clients often specify a design philosophy in their documentation, and individual designers and companies may have their own in-house approaches. It is good practice for a formal design philosophy to be written as one of the first documents on a design project. Similarly a safety and loss prevention philosophy is ideally produced early on in the design process.

The design philosophy should record the standards and philosophies used, together with underlying assumptions and justifications for the choice. This is both to allow a basis for checking in the detailed design stage and for legal purposes.

In the absence of written philosophies, a second engineer at detailed design stage might attempt to apply the ones he or she would have chosen, and the plant may become subject to pointless expensive and extensive redesign.

SPECIFICATION

There are a number of types of specifications which are produced or introduced at various stages of the design process. We might split them initially into specific and boilerplate categories. Let us first dispense with the second category.

Boilerplate is a term used in legal contexts, but it originates in engineering. A rating-plate containing standard text was required to be attached to a boiler by the Boiler Explosions Act (1882), and the term subsequently came to mean standard text which is cut and pasted into documents.

Much of this sort of text has been described as "write only" documentation (or WOD—we engineers love a three-letter acronym (TLA)), because someone has to write it but virtually no one has to read it. The consultant draws the attention of the designer to it, and they in turn draw it to the attention of their suppliers.

The tender documentation sent out to contractors by consultants frequently contains great volumes of boilerplate specification, lists of applicable (or potentially inapplicable if your consultant is lazy and/or risk averse) legislation and standards, and references to all the other things which the conceptual designers didn't personally consider, but think someone else should. I am not convinced that this frequently lazy approach actually provides the degree of legal protection those responsible for it imagine.

More usefully, there will be a far thinner volume of specifications which inform the definition of the design envelope. The expected quantities and qualities of feeds into the process should be included, as well as a description of product quality and quantity. These descriptions will ideally be in the form of ranges of concentrations, flows, temperatures, pressures, and so on. There may be statistical information to allow the designer to understand the distribution of likelihood of various conditions.

There may be reference to specifically applicable standards, legislation and so on. This differs from boilerplate as those responsible for the previous stage of design have identified that these documents are likely to really matter to this specific design.

The separation of these useful specifications from the boilerplate is often the first job of the contractor's plant design engineers. The boilerplate has to be checked for anomalous content alongside the real design process, but that is not usually on the critical path. It usually suffices to send out (largely unread) relevant sections to those offering prices for the equipment to be purchased.

PROCESS FLOW DIAGRAM (PFD)

The Process Flow Diagram (PFD) is a visual representation of the mass and energy balance. The PFD treats unit operations more simply than the Piping and Instrumentation Diagram (P&ID—see next section). Unit operations are shown using British Standard (BS) P&ID symbols or sometimes as simple blocks, pumps are shown, as are main instruments (Figure 3.1).

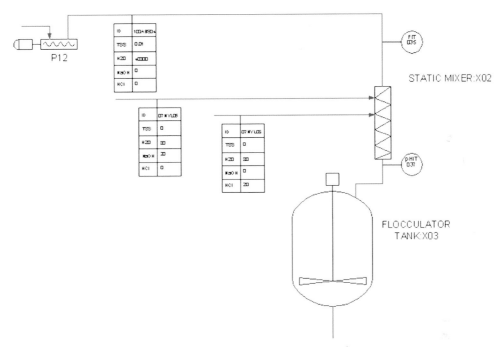

Figure 3.1 PFD for the pH correction section of a water treatment plant.

The lines on this diagram are labeled in such a way as to summarize the mass and energy balance, with flows, temperatures, and compositions of streams.

Please do not call this a flow sheet, as this term is used to mean quite a few different things (for my suggested use of the term in a process plant design context see Chapter 14). Neither should you think that a simulation program printout is a substitute for a professional PFD conforming to a recognized standard.

The Block Flow Diagram (BFD) used in academia as a simplified substitute for a PFD is not something I have seen used in practice, other than when drawn on a beermat in a pub discussion.

The general British Standard for engineering drawings, BS 5070 applies to the PFD, as well as BS EN ISO 10628. The symbols used on the PFD should ideally be taken from BS EN ISO 10628, BS1646, and BS1553.

PIPING AND INSTRUMENTATION DIAGRAM

The P&ID is a drawing which shows all instrumentation, unit operations, valves, process piping (connections, size, and materials), flow direction, and line size changes both symbolically and topographically (Figure 3.2). Thus it is not a scale drawing, and the lines on a P&ID turning corners mean nothing, though the joining of three or more streams is meaningful.

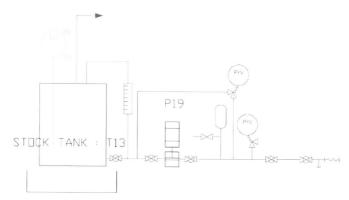

Figure 3.2 Extract from a P&ID.

The P&ID is the process engineer's signature document, and its purpose is to show the physical and logical flows and interconnections of the proposed system. Recording them visually on the P&ID allows them to be discussed with software engineers, as well as other process engineers. There are a great many variants in additional features between industries, companies, and countries, but producing the drawing to a recognized standard makes it an unambiguous record of design intent, as well as a design development tool. This is, however, rather idealist. I have only ever worked for one client who produced unmodified BS P&IDs.

The standards for the symbols which should ideally be used by British engineers are BS EN ISO 10628, BS1646, and BS1553 and, like all engineering drawings, it should be compliant with BS 5070. Having said how things should be, how things actually are is that many companies and industries have their own internal standards for P&IDs. Professional engineers get used to the range of symbols and conventions commonly used on P&IDs (though I try to shield my students from this at first to avoid confusion).

There are also a number of P&ID conventions which do not appear in standards:
- Flow comes in on the top left of the drawing, and goes out at bottom right.
- Process lines are straight and either horizontal or vertical.
- Flow direction is marked on lines with an arrow.
- Flow proceeds ideally from left to right, and pumps, etc. are also shown with flow running left to right.

- Sizes of symbols bear some relation to their physical sizes: valves are smaller than pumps, which are smaller than vessels, and the drawn sizes of symbols reflect this.
- Unit operations are tagged and labeled.
- Symbols are shown correctly orientated: vertical vessels are shown as vertical, etc.
- Entries and exits to tanks connect to the correct part of the symbol—top entries at the top of the symbol, etc.

Less complex P&IDs produced during earlier design stages will normally come on a small number of ISO A1 or A0 drawings, but the P&IDs for a complex plant may be printed in the form of a number of bound volumes where every page carries a small P&ID section.

Every process line on the drawing should be tagged in such a way that its size, material of construction, and contents can be identified thus:

Number showing NB in millimeters—Letter code for material of construction—Unique line number—Letter code for contents e.g. a line tagged 150ABS004CA would be a 150 mm NB line made of ABS (plastic), numbered 4, containing compressed air. Personally, in common with many other designers, I number the main process line components first, increasing from plant inlet to outlet. So line 100ABS001CA would, for example, normally be upstream of line 150ABS004CA above. Design development can, however, mean that this gets a bit muddled on the as-built version of the drawing.

Every valve and unit operation on the P&ID will also be tagged with a unique code; a common key is given in Table 3.1:

Table 3.1 P&ID Tag Table
Valves

MV	Manual valve
AV	Actuated valve
FCV	Flow control valve
CV	Control valve
ESV	Emergency shutdown valve

Unit operations

U	Unit
T	Tank
P	Pump
B	Blower
C	Compressor

The letter code will be followed by a unique number for that coded item.

Similarly, every instrument will be given a code as set out in BS1646 as follows:

First Letter—measured parameter:

L = Level

P = Pressure

T = Temperature

Additional letters—what is done with the measurement (you can have more than one of these):

I = Indicator

T = Transmitter

C = Controller

The letter code will be followed by a unique number for that coded item, for example, PIT1 would normally be the first pressure indicator/transmitter on the main process line.

The British Standards cover these conventions in more detail.

The P&ID is a master document for Hazard and Operability (HAZOP) studies. It also frequently shows useful termination points between vendor and main contractor, and between main contractor and equipment supplier.

FUNCTIONAL DESIGN SPECIFICATION (FDS)

An FDS is sometimes called a control philosophy, although both of these terms are used in other contexts to mean other things. The document I am referring to describes (ultimately, in practice, for the benefit of the software author) what the process engineer wants the control system to do.

It starts with an overview of the purpose of the plant and proceeds to document, one control loop at a time, how the system should respond to various instrument states, including failure states.

This is done in clear and straightforward language, designed to be entirely unambiguous and comprehensible by nonspecialists.

It is read in conjunction with the P&ID and refers to P&ID components by tag number, and is used alongside the P&ID in HAZOP studies.

PLOT PLAN/GENERAL ARRANGEMENT/LAYOUT DRAWING

The General Arrangement (GA) drawing shows the plant and pipework as it is intended to be installed (or as it was installed in the case of an "as-built" GA).

In professional practice a specialist piping or mechanical engineer may produce the finished drawing, but chemical engineers lay out equipment in space, and produce this drawing as part of their design process (Figure 3.3).

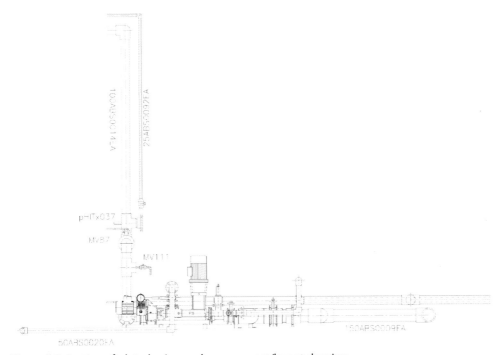

Figure 3.3 Section of plot plan/general arrangement/layout drawing.

Drawings should conform to BS5070 and show (as a minimum), to scale, a plan and elevation of all mechanical equipment, pipework, and valves which form part of the design, laid out in space as intended by the designer. Where possible, the tag numbers used in the P&ID should be marked on to their corresponding items on the GA as well, to allow cross-referencing.

The inclusion of key electrical and civil engineering details is normal in professional versions, and there are also usually detailed versions for each discipline which refer back to a common master GA.

Ideally, the drawing will be produced to a commonly used scale (1:100 being the commonest scale for these drawings), and would be marked with weights of main plant items. Fractional or odd-numbered scale factors should be avoided. Sectional views demonstrating important design features are a desirable optional extra.

PROGRAM

A project "program" or "schedule" most usually refers to a Gantt chart which sets out the planned timescale and resourcing of a project. While a specialist may produce this in a larger company, chemical engineers should be able to produce a competent program as part of the plant design process. In the absence of a resourced program (Figure 3.4), any estimate of capital cost must be treated with great suspicion.

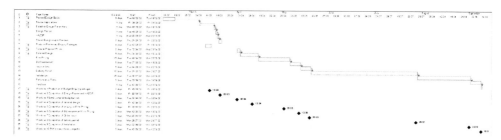

Figure 3.4 Example Gantt chart.

The overall project is broken down into discrete tasks, and the time and resources necessary to achieve each of these tasks is estimated. "Milestones" usually appear at the end of phases or tasks, and are often associated with the production of deliverables, triggering payment or another phase of the project. "Dependencies" have to be identified—certain tasks must precede or follow others. Additional time ("float" or "bunce") should be added to the minimum reasonable times for each task, to reflect the uncertainty of the estimate.

A program can then be generated which shows a reasonable estimate of the time to complete all tasks, which can be analyzed to see which activities set overall project time. The critical path through a program involves the chain of activities whose completion is critical to overall program length.

Microsoft market a programming tool called "Project" which can be used to produce this document to a professional looking standard, though in practice, other specialist tools such as "Primavera" are at least as commonly used by specialist project programmers.

COST ESTIMATE

Academic approach

Cost estimation is usually taught in academia using the Main Plant Items/Factorial method, a method I have never once seen used in professional practice, though the prices it produces are normally acceptable as very broad ballpark estimates.

It is well explained in Sinnot and Towler (aka Coulson and Richardson 6), a book all chemical engineers should be familiar with, so I will not reproduce it in detail. In outline, however, student-style pricing goes like this:

- Price up your main unit operations, probably using curves of unit price against duty.
- Use factors which accompany curves to account for quality of materials, pressure, etc.
- Multiply prices by the CPI (Chemical Price Index) or similar to reflect sector specific inflation since curves were drawn.
- Convert prices obtained to desired currency at today's prices.
- Add together all the prices.
- Multiply this total by Lang factors to estimate cost of all the other items and services necessary to get a complete plant (around 4).
- Marvel at how big the numbers are (even though they are probably a radical underestimate).

The technique does, however, raise a number of important issues, and gives a feel for the relationships between the prices of the inputs to a project.

Students come away with the useful impression that the cost of a complete plant is a great deal more than the cost of delivering its main unit operations to site, and that electrical and civil costs in particular are very significant. It also introduces students to the idea that plants are built to make a profit—one of the Lang factors even has that name.

It is, however, usually insufficiently powerful to genuinely resolve differences between options—the margin of error is probably more like $\pm 50\%$ than the $\pm 10\%$ many of my students think it is.

This is, however, incredibly rigorous in contrast to the version of the "Economic Potential" technique I have seen used by academics, who just want to get to the pinch analysis as soon as possible, and don't wish trivia like safety, cost, and robustness to get in the way.

Their technique is as follows:

- Google the bulk price of the proposed feedstock (F) and expected product (P).
- If P > F, any process which turns F into P is economic.

The former technique might be a bit woolly, but the latter is operating at an accuracy of $\pm$ several hundred percent. It is not merely worthless; its use encourages overly complex and uneconomic design choices. It does, however, have the advantage of taking only a few seconds.

Professional budget pricing

Even the least rigorous real-world pricing exercises tend to start from quoted prices for main plant items. If you work in a contracting company, reasonably recent quotes for reasonably similar equipment will usually be available.

Costs of control panels, software, electrical and mechanical installation, civil and building works will also be priced based upon past quotations or rules of thumb.

Internal cost will be estimated based on experience and/or rules of thumb. A good chunk of contingency will be added to reflect the high degree of uncertainty at this early stage of the job.

Someone who does this for a living will be able to produce a $\pm 30\%$ budget price in this way in a few hours.

Professional firm pricing

If a company is going to contract to build a plant for a fixed sum of money, it needs to be certain that it can make a profit at the quoted price.

Equipment prices are obtained from multiple sources for specific items whose specifications come from reasonably detailed design.

Civil, electrical, and mechanical equipment suppliers and installation contractors are also invited to tender for their part of the contract.

Internal quotations are also usually obtained from discipline heads within the company for the internal costs of project management, commissioning, and detailed design.

There may well be negotiations with all of these sources of information. Ideally there will be multiple options for equipment supply and construction. A price based on a single quotation is far less robust than one which has a broader base.

Once there are prices for all parts and labor, residual risks are priced in. The profit, insurances, process guarantees, defect liability periods, and so on are then added.

This exercise can take a good-sized team of people weeks or months to complete, and the product is a $\pm 1-5\%$ cost estimate.

There was a recent article in *The Chemical Engineer* magazine about a company using an Excel spreadsheet they had developed to produce Class 3 budget price estimates at an early stage by a process they call "conceptual design emulation." I have no idea how well this works, but do I know it isn't free, although the conventional approach is far from cheap.

EQUIPMENT LIST/SCHEDULE

A schedule or table of all the equipment which makes up the plant is usually first produced at FEED stage. Tag numbers from drawings are used as unique identifiers, and a description of each item accompanies them. There may be cross-referencing to P&IDs, datasheets, or other schedules.

Similar schedules are produced for all instrumentation, electrical drives, valves, and lines (Figure 3.5).

INSTRUMENT SCHEDULE

								Rev 0	Rev 1	Rev 2	Rev 3	Rev 4	
Project	Permanent Effluent/Groundwater Treatment Plant	Project Ref			Prepared by	SMM							
Project Site		Document Ref			Checked by	STM						CONFIRM	
Client					Date	30/04/2004							
Client Ref					Approved by								
					Date								

Inst No	Description	Supplier	Type	P&ID no	Line No	Size (mm)	Material of Constr	Design Fluid conditions		Operating range		Alarm conditions		Notes	CSL /Equip supplier
								Press (bar)	Temp (C)	Min	Max	H	L		
LC001	MH 102 (PS1) Level Controller	Milltronics	Ultrasonic	90501 002		n/a	proprietary	atmos	ambient	0	6000 mm	Y	Y	Milltronics Multiranger with 2 sensors in sump panel mounted indicator Some alarms from PLC	CSL
PI002	Pressure Indicator	TBA	Bourdon	90501 002	0056	tbc	SS enclosure	5	ambient	0	2 bar	N	N	Standard 50-75/100 mm pressure gauge	CSL
PTx003	Pressure Transmitter	GEMS	Transducer	90501 002	0056	n/a	316 ss	6	ambient	0	2 bar	Y	Y	Pressure transmitter and panel mounted indicator	CSL
FTx004	Flow Transmitter	ABB	Electromagnetic	90501 002	0056	300	proprietary	16	ambient	0	120 m3/hr	Y	Y	AB Magmaster inline meter with panel mounted display	CSL
LC005	MH 92A (PS2) Level Controller	Milltronics	Ultrasonic	90501 002		n/a	proprietary	atmos	ambient	0	6000 mm	Y	Y	Milltronics Multiranger with 2 sensors in sump panel mounted indicator Some alarms from PLC	CSL
PI006	Pressure Indicator	TBA	Bourdon	90501 002	0062	tbc	SS enclosure	5	ambient	0	2 bar	N	N	Standard 50-75/100 mm pressure gauge	CSL
PTx007	Pressure Transmitter	GEMS	Transducer	90501 002	0062	n/a	316 ss	6	ambient	0	2 bar	Y	Y	Pressure transmitter and panel mounted indicator	CSL
FTx008	Flow Transmitter	ABB	Electromagnetic	90501 002	0062	300	proprietary	16	ambient	0	240 m3/hr	Y	Y	AB Magmaster inline meter with panel mounted display	CSL
LC009	MH 55A (PS3) Level Controller	Milltronics	Ultrasonic	90501 002		n/a	proprietary	atmos	ambient	0	6000 mm	Y	Y	Milltronics Multiranger with 2 sensors in sump panel mounted indicator Some alarms from PLC	CSL
PI010	Pressure Indicator	TBA	Bourdon	90501 002	0066	tbc	SS enclosure	5	ambient	0	2 bar	N	N	Standard 50-75/100 mm pressure gauge	CSL
PTx011	Pressure Transmitter	GEMS	Transducer	90501 002	0066	n/a	316 ss	6	ambient	0	2 bar	Y	Y	Pressure transmitter and panel mounted indicator	CSL
FTx012	Flow Transmitter	ABB	Electromagnetic	90501 002	0066	300	proprietary	16	ambient	0	540 m3/hr	Y	Y	AB Magmaster inline meter with panel mounted display	CSL

Figure 3.5 Extract from an equipment schedule.

Some modern software promises to remove the necessarily onerous task of producing these schedules from the junior engineer's tasklist. The nit-picking rigor required is certainly arguably more suited to the infinitely patient stupidity of computers than the inventive mind of a professional engineer, but schedules are still mostly generated by unlucky people.

DATASHEETS

Datasheets gather together all the pertinent information for an item of equipment, mainly so that nontechnical staff can purchase it. Process operating conditions, materials of construction, duty points, and so on are brought together into this document to explain to a vendor what is required (Figure 3.6).

Equipment Schedule and Pricing

Supplier

Client		Prepared by	STM		Date	Nov-03	
Site		Checked by			Date		
Project	Example	Contract ref		C2265_C2295	Rev	0	
Section	Skid mounted pumps	Section	X	of	Y		REV
Compressor					Item	F16,F17	

One air compressor, pressurised air storage reservoir and associated controls and ancillaries to provide air for oil water separator (if required), continuous sand filter (if required), DAF unit (if required), air actuated valves and instruments. Current duty is estimated from preliminary calculations. Suppliers should satisfy themselves that the duty specified is correct and make any adjustments as necessary to the sizing. Supply should be for complete free standing unit. Dryer to be provided if supplier considers this to be beneficial. System will be located inside control building.

Any information not requested which the supplier believes will add to his offer to be provided on separate sheets

Perfomance parameter	Description	Pipework	Valves	Option1	Option 2	Option 3	Option 4
Min flow (Nm3/hr)				30	10	10	10
Max flow (Nm3/hr)				50	20	25	35
Pressure (bar)				[illegible]	[illegible]	[illegible]	[illegible]
Compressor							
Manufacturer							
Model							
Type							
Materials of construction							
Noise levels (dBA)							
Length (mm)							
Width (mm)							
Height (mm)							
Weight (kg)							
Inlet connection (mm dia)							
Outlet connection (mm dia)							
Motor type							
Enclosure class (IEC 34-5)							
Insulation Class (IEC 85)							
Motor Speed (rpm)							
Rated power (kW)							
Efficiency at design duty (%)							
Mains frequency (Hz)	50						
Rated voltage	415						
Rated current (amp)							
Starting (Star Delta/DOL)							
Starting current (amp)							
Price for complete system, delivered Manchester							
Delivery period for complete, tested system (weeks)		Target <10 weeks					

Enter here any additional notes, details of additional equipment or facilities required

Figure 3.6 Example of an equipment datasheet.

Data sheets need to be cross-checked with a number of drawings, calculations, and schedules, and care has to be taken to ensure that they are in accordance with the latest revisions. This is a more skilled task than the generation of schedules, and will therefore be likely to remain in the purview of young engineers for years to come.

SAFETY DOCUMENTATION

HAZOP study

As I and all professional engineers use the term, a HAZOP study is a "what-if" safety study. It requires as a minimum a P&ID, FDS, process design calculations, and information on the specification of unit operations, pumps, etc. as well as probably eight professional engineers from a number of disciplines.

In an academic setting, the calculations and specifications of equipment may stand in for the datasheets which would be available to a real HAZOP.

The report produced by the participants will usually nowadays include a full description of the line-by-line (or node-by-node) permutation of keywords and properties used in carrying out such a study, but it was more usual in the past to produce a summary document listing only those items which were identified by HAZOP as being problematic, what the problems were and how it was intended that they be avoided.

In today's litigious environment, a full and permanent record of all that was discussed and considered in a HAZOP is increasingly considered prudent. This may most conveniently be achieved by video recording of the entire procedure. Recording the entire procedure one way or another is now considered best practice.

As students frequently have a great deal of difficulty imagining what they might do about any problematic upset conditions they have identified, I have included at Appendix 2 an upset conditions table from "Process Plant Design and Operation—Guidance to Safe Practice," by Scott & Crawley which offers useful guidance.

Zoning study/hazardous area classification

Zoning the plant with respect to the potential for explosive atmospheres is not a strictly quantitative exercise (Figure 3.7).

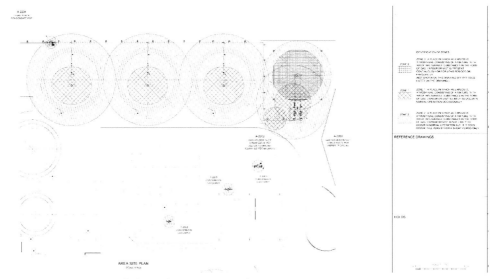

Figure 3.7 Zoning study/hazardous area classification. *Copyright image reproduced courtesy of Doosan Enpure Ltd.*

It is common for a small number of engineers to get together with design drawings to produce a zoning drawing or drawings showing the explosive atmosphere zoning they think appropriate for the various parts of the plant. There are more details of this in the chapter on layout.

DESIGN CALCULATIONS

These are usually not really a deliverable at all, since only other process engineers can understand them. Design proceeds by a number of stages, from initial coarse approximations to the level of fineness specified in the design brief issued. Academic exercises are normally carried out around the level of detail used for budget costing (though they often use tools which would not be used at this stage in practice).

If the design is not utterly novel, heuristic design is realistic for the majority of items. First principles design should not be favored for items which are commercially available, because this is unrealistic, is the opposite of professional practice (and, in a teaching environment, it is difficult to detect cheating). If it is desired to evaluate students' ability at first principles design, a genuinely completely novel item should be chosen.

Process designs normally aim to determine certain key dimensions, areas, and volumes based on a number of parameters. For nonnovel processes there may be rough rules of thumb, established design guides, standards, or codes of practice.

In professional practice these will be combined to inform a design sizing. Only in the near-complete absence of relevant guidance of this sort will first principles design be used, and large design margins will be added to reflect the high degree of uncertainty.

More generally, an understanding of the degree of applicability of the design methods used and uncertainly around design data should be reflected in a stated added design margin.

The design calculations should include consideration of construction, materials selection, cost, and practicality, and should cover all foreseeable operating conditions, including start-up, shutdown, and maintenance, as well as environmental considerations such as power outage, plausible natural and man-made disasters. The range of variability in feed stock flows and compositions under normal conditions should be considered as well as these more extreme variations from steady state.

Opaque outputs from modeling and simulation programs are no substitute for transparent process design calculations, as it is easily likely that neither the author nor the checker can readily understand what is going on in the simulation.

In assessing process design calculations it should be borne in mind that the point of such calculations is not to demonstrate proficiency in mathematics or chemistry. The point of process design calculations is to produce a minimum specification for an item, so that the next size up (or sometimes down) commercially available unit can be purchased with a reasonable degree of confidence in its robustness.

Process design is not a matter of finding exactly the right size item, but of finding one large and flexible enough such that one can be sure that the commissioning engineer can make it work.

Complementary with the process design calculations are the calculations used to size pumps, pipework, channels, and so on, which we might collectively refer to as hydraulic calculations.

Precise determination of dynamic heads by fluid mechanics is extremely difficult in practice and, furthermore, completely unnecessary. There are a number of heuristics which may be used to carry out rapid determinations of approximate headlosses to distinguish between competing conceptual designs, and produce initial layouts. The use of appropriate tabulated values and nomograms for this purpose should be entirely acceptable for early stage design.

More detailed calculations should be required for the final selected design. These should at a minimum degree of rigor be based on one of the friction factor methods, with k-values for fixtures and fittings. All dimensions used in the calculations should refer back to the dimensioned GA drawing or "iso". The calculations should be based on actual selected commercially available pipework, valves, pumps, instrumentation, and so on. (For those teaching this subject, the internet means that our students will no longer need to bother manufacturers to obtain this data, so we have no reason not to require this degree of realism.)

A fairly common approach (and my personal preference) is to use one tab of a spreadsheet program for each unit operation, and to link the inputs and outputs to the tabs in such a way that the whole spreadsheet is a standalone model encompassing mass and energy balance, unit operation sizing, and hydraulic calculations. I use a standard template which looks like an engineer's calculation pad. Each tab has a vertical column of these virtual pages in which the argument and calculations for that unit operation is set down in a logical and readable form (Figure 3.8).

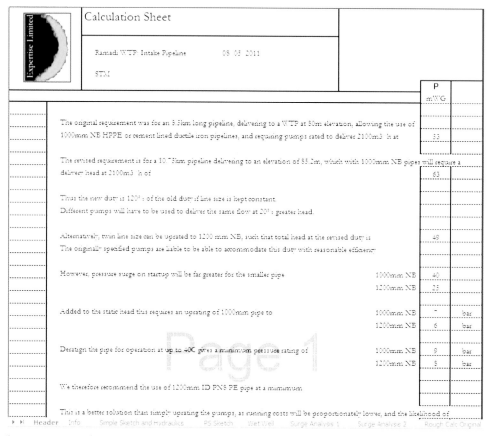

Figure 3.8 Example intake pipeline design calculations.

However it is achieved, transparency and clarity of design intent should be the important factor in evaluating process and hydraulic design calculations. Calculations should be double-spaced, ideally on an engineers' calculation pad or electronic facsimile, and every step should refer to any drawings, design standards or other references upon which it relies.

ISOMETRIC PIPING DRAWINGS

At the detailed design stage, piping isometrics are produced for larger pipework, either by hand on "iso pads" or by computer-aided design (CAD) (Figure 3.9):

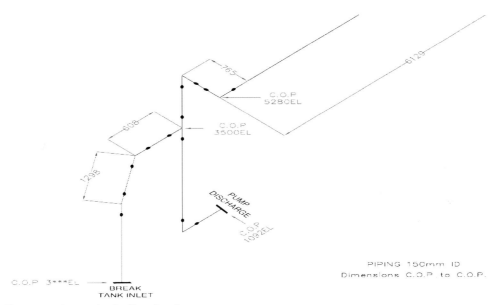

Figure 3.9 Isometric pipeline drawing.

Isometric piping drawings are not scale drawings, so they are dimensioned. They are not realistic, pipes are single lines, and symbols are used to represent pipe fittings, valves, pipe gradients, welds, etc. Lines, valves, etc. are tagged with the same codes used on the P&ID and GA. Process conditions like temperature, pressure, and so on may also be put on the iso.

It may well be that "clashes" where more than one pipe or piece of equipment occupy the same space are only identified at the stage of production of isos, so design cannot be considered complete before isos are produced.

The purpose of the iso is to facilitate shop fabrication and/or site construction. They are also used for costing exercises and stress analysis, as they conveniently carry all the necessary information on a single drawing.

Producing isos by hand is quite time-consuming. There are some CAD systems now which can automatically produce isometric drawings from the GA drawings.

It is claimed that these reduce drafting errors and inconsistencies, spot clashes earlier, facilitate links to other software such as costing programs, and save time. Hand drafting is, however, still the norm in many industries.

SIMULATOR OUTPUT

This is not really a deliverable at all, as at best only other process engineers can understand it. The purpose of modeling and simulation on those few occasions where it might be used by process designers is to provide supporting information.

I have also seen simulator outputs used to attempt to resolve conflicts between engineers as to which heuristics are most applicable, but in such circumstances it becomes very clear that the output of such programs has a great deal to do with who is choosing the inputs.

If both engineers are competent, the use of a model in my experience merely makes clear that they are arguing about which design basis or heuristic is most applicable to the situation in question.

FURTHER READING

Anon 1977. Specification for graphical symbols for general engineering. Piping systems and plant. BS 1553-1:1977 BSI Standards.

Anon 1988. Engineering diagram drawing practice. Recommendations for general principles. BS 5070-1:1988 BSI Standards.

Anon 1999. Symbolic representation for process measurement control functions. BS 1646 BSI Standards.

ISO, 2001. BS EN ISO 10628:2001. Flow diagrams for process plants—General rules. BSI Standards.

Kauders, P., 2014. Plant Design. How Much? The Chemical Engineer, London.

Kletz, T., 1999. Hazop and Hazan. Institution of Chemical Engineers, London.

Sinnot, R.K., Towler, G., 2005. Chemical Engineering Design, Vol. 6. Butterworth-Heinemann, London.

CHAPTER 4

Twenty-First Century Process Plant Design Tools

GENERAL

There is a far greater use of computers in process plant design than when I started in practice. Hand drafting of most drawings is no longer really practiced, and handwritten calculations are also uncommon. Neither do we need to write our own computer programs as I had to if I wanted a computer to do something for me back in 1990.

There is also some use of modeling and simulation programs to support design activity, especially in certain sectors. This has not, however, replaced design activity, and it cannot, for reasons I will explain in this chapter.

Professional plant design engineers worldwide and across sectors only use a small subset of the available programs, largely for reasons of economy and consistency to allow information sharing.

We more or less all use MS Excel, MS Project, and Autodesk AutoCAD. Modeling and simulation programs tend to be sector-specific specialist products. Oil and gas industry specific modeling and simulation programs tend to be popular in universities, but there are equivalent specialist programs (such as the Hydromantis products used in my sector) which never seem to feature in university courses.

Researchers in chemical engineering departments make a lot of use of simulation and modeling programs in their research, and they have consequently started making use of these programs in teaching the thing they call "design." Some of these programs (such as, e.g., Matlab) are never used by professional engineers, and are being shoehorned into a duty they are unsuited for. Some (notably Hysys) are frequently misused by researchers unfamiliar with professional practice to fill a gap in their own knowledge.

These programs are highly discounted or even free to academia, and are often incredibly expensive to commercial users. Thus, many of the computer programs used to teach the thing known as "process design" in academia may be irrelevant, misused, or financially nonviable for commercial use.

In this chapter I will discuss a few of the programs actually used by plant designers, as well as the more popular programs misused by researchers for "design" teaching.

Before I do that, I would like to draw your attention again to how we ought to be using computers, according to the Institution of Chemical Engineer's (IChemE's) Computer-Aided Process Engineering (CAPE) subject group.

USE OF COMPUTERS BY CHEMICAL ENGINEERS

All the new tools used by chemical engineers are computer based, and the IChemE guidelines on use of computers by chemical engineers should be followed. The most important thing to understand about these design tools is that they are intended to support, rather than replace, professional judgment. The guidelines summarize themselves with the following key points:

> Management has the overall responsibility for developing appropriate standard procedures and practices and for ensuring that they are followed.
> It is a professional engineer's legal and professional responsibility to exercise good engineering judgment in making design decisions and, therefore, to satisfy him/herself regarding the adequacy of the information upon which design decisions are based. This means you!
> Much of this information is today generated by computer-based systems and so the quality of these systems and the skill and judgment with which they are applied to a design problem are a critical part of these responsibilities.
> The purpose of these Guidelines is to suggest some simple precautions which should be taken to help protect the integrity of proposed engineering solutions and thus to adequately discharge professional responsibilities, for example:
> What matters is the quality of the engineering decision: focus on "fitness for purpose" of both the computer-based system and the data which is fed into it
> Assume that everything is "guilty until proven innocent": you must check and ensure that the computer-based model is appropriate to your needs and that the data (including any data from databanks, etc.) is correctly specified and adequately covers the expected ranges (for example, of temperatures, pressures and compositions).
> You must check and ensure that the program has worked successfully and that the results are adequate for your purpose: you must satisfy yourself that you fully understand any weaknesses and that you apply them sensibly and with good engineering judgment
> Sensitivity analysis is a key weapon in identifying where the critical problems lie and in assessing their likely impact on your design decisions.
> Program development is not a trivial job and to do it well requires special skills and experience.
> Engineering decisions will be based upon the results generated by these programs. The program must therefore work correctly and proper records must always be kept.
> Do not hesitate to seek help and guidance from your more experienced colleagues, from your support services or even from the suppliers of the systems concerned (and seek it early, not when things have already gone wrong).

Unfortunately, these principles seem not to be commonly understood. Such programs are consequently being used to carry out tasks they were not intended to be used for. Worse still, such misuse is sometimes actively taught in universities as proper practice, and engineering employers consequently have to actively reeducate their graduate intake.

IMPLICATIONS OF MODERN DESIGN TOOLS

The workhorse programs allow for great increases in productivity and the possibility of more decentralized and flexible engineering services. When I graduated, calculations were done by hand on paper pads by engineers. Drawings were exchanged between offices in hard copy by courier. Copying of drawings was done by means of a machine which produced the blue lines on beige background known as a "blueprint" (and a terrible smell of ammonia). Fax might be used to transmit A4 copy. There might be one PC shared between a dozen engineers, and time had to be booked on it. If you wanted a program, you needed to write it yourself, so it tended to be a bit buggy.

Now, an engineer working alone at a PC can research alternatives, carry out his own calculations with reliable and extensively debugged programs, generate his own drawings on his computer, and collaborate with others worldwide using more or less instantaneous communications. He can send and receive editable versions of his drawings to and from Australia in seconds.

This universal use of computer-aided design (CAD) and web-based communications has also had the effect of closing the drawing offices which were a feature of engineering companies back when I started. There are now drawing offices in India and elsewhere which will produce your drawings for you at a fraction of UK drafting rates, but even they rarely feature drawing boards.

The general implication of all modern design tools is that they can harness a great deal of stupid, patient computing power. This can be used to produce models of process plants in Excel, dedicated modeling and simulation programs, or even programs like Matlab which can be used to throw ourselves back to the time when we had to write our own programs.

So we can use computers to produce reliable, and transparent models, or models which are rather opaque to all but the most experienced engineers. Just as banks will only lend money to people who don't need it, the only people who should use many modern design tools are those who don't need them.

Both lecturers and students often think that because modeling and simulation programs work on the basis of first principles, their output is somehow more rigorous or better understood than heuristic design. I am, however, fairly sure that even those who write such programs do not completely understand them, as no one can completely understand a computer program beyond a fairly low level of complexity. I am also fairly sure that those who write these programs would not understand the plant "design" produced by such programs, even if they did understand the program itself, for similar reasons of complexity.

Those who think that the output of such programs is more trustworthy or transparent than professional approaches do not understand that all approaches are approximate and heuristic, but that the professional approach is based on producing a model simple enough for complete human comprehension, founded as directly as possible in empirical evidence and professional judgment, and tested at full scale by generations of professional process designers.

The modeling–as–design approach on the other hand is based in a necessarily cut-down version of pure theory, running on a desktop computer in a model which no one fully understands, written by people without any experience in designing process plants, untested at any scale, and never intended to be used as a substitute for real design calculations.

CATEGORIES OF DESIGN

Unit operation sizing and selection

Universities still teach students approaches to unit operation sizing based on hand calculations using charts which come from the slide-rule era. Many who do this think it helps students to understand the unit operation—whether this is true or not, it isn't much to do with modern professional practice, since process plant designers mostly use spreadsheet programs to do their calculations.

Equipment suppliers are the ones who do detailed calculations to specify unit operations (or more likely punch numbers into a proprietary spreadsheet or program), so that what goes on in universities is at best a relic of an earlier era.

Since spreadsheet programs can't read charts, many of these techniques aren't of much practical use any more. We need equations rather than charts to allow us to work with spreadsheets.

Modeling and simulation programs are also used to "design" unit operations, in as much as they are used to model a number of unit operations with a view to writing specifications for the equipment.

This kind of "design" is far more common in operating companies than in the contracting companies or design houses who design complete plants and offer process guarantees for them. Even in operating companies, simulation programs are supposed to support design activity rather than substitute for it.

Operating companies have access to large quantities of real plant-specific data, so they can tune and validate the simulation programs to make them match their plant. Whole-plant designers do not have this luxury, as their plants have not yet been built, and they lose access to operating data for the plants they do build once commissioning is completed.

Hydraulic design

In the hydraulic calculations used to size pumps and pipework, empirical approximations like the Colebrook–White have superseded the Moody diagram, for the reasons discussed in the last section—MS Excel can't read diagrams.

The modeling and simulation programs used for plant "design" can usually do some hydraulic calculations, but often come with a set of defaults which causes problems for the unwary or uncritical user.

There are quite a number of fluid dynamics modeling programs which can be used for complex fluid dynamics problems, but these are a bit overpowered and slow for the most common hydraulic calculations.

Mass balance

I would expect that anyone reading this book would know what this is but, just in case, process plant designers work out how much stuff is going to flow from one place to another in their plants by applying a simple principle—if matter is neither created nor destroyed, all the masses of stuff must add up.

This "mass balancing" is usually done by practitioners using MS Excel. It is particularly common and helpful to have one unit operation per tab, and have the mass balance expressed in making links between tabs. My suggested methodology is explained in Chapter 13.

Energy balance

Energy is not created or destroyed either, so we can work out the energy needs and yields of our plant by balancing energy inputs and outputs.

The energy balance is usually built into the same Excel spreadsheet as the mass balance.

TOOLS—HARDWARE

Mobile devices

Price £1—500

More or less everyone has got one of these with them all the time nowadays, and they all have a basic calculator built in, which is perhaps why no one can do mental arithmetic any more.

You might need to have pi and a few other numbers like root 2 and root 3 memorized to a few decimal places (or you can google them with your mobile in the case of my students), but I find that the calculator can do most of my back-of-an-envelope calculations.

Many can also run Excel-compatible spreadsheet programs for when you want to do something a little trickier, and have a permanent record of your calculations which you can tweak up in Excel later.

When I asked my students to evaluate graphical calculators, one asked me why they would pay £200 for one when everybody already has a mobile device which can do pretty much everything it can, as well as many other things as well.

Handheld calculators

Price £10–20

These are the slide rules of the modern era, but mine doesn't make it out of my desk drawer very often nowadays.

Sums I can't do in my head are more likely to be done with my mobile phone than with my far more powerful scientific calculator. Anything remotely tricky gets done in a spreadsheet, which allows for easy checking, and automatically records the output of the calculation.

I have a friend who is an enthusiast for the graphical calculators now commonly available, but usually banned from school and university examinations in a way which limits their use in academia.

Graphical calculators

Price £100–200

My enthusiastic friend tells me that modern graphical calculators can not only produce full-color graphs, they can also do calculus.

I guess they might be very handy once you have mastered their operation, but they are ultimately a kind of handheld calculator and there are reasons why my conventional one doesn't see much daylight.

The use of a spreadsheet program on a PC automatically produces a checkable, annotated, and fairly permanent record of the calculations you have carried out, and the assumptions made in doing so. Quality Assurance (QA) and traceability of documentation is very important in professional practice.

It turns out that these calculators are easily linked to a PC and import and export commonly used file formats. I looked into a couple of models which were recommended to me: the TI Nspire and the Casio FXCG20.

The TI calculator solves differential equations, which the Casio unit does not, but TI have pulled these calculators from the UK market due to lack of interest, so it is hard to get hold of one.

Both look like they might be pretty good teaching tools, as they come with an emulator which runs on a PC which you can use to demonstrate their use. I'm not sure what advantages they would give to a practitioner though.

TOOLS—SOFTWARE

PCs do most of the heavy lifting nowadays in process design. Engineers have to care about the price of products so, in the course of researching this book, I attempted to obtain prices for all the programs I looked at.

This was trickier than I thought it would be and, consequently, many of the prices quoted in the following sections are approximate. There seems to be a culture of secrecy about pricing in the sector, and the generation of confusion with complex price lists.

I think, however, that the prices quoted are approximately correct for similar functionality.

Spreadsheets

MS Excel

Single user license £110.

This is the workhorse which does most chemical engineering calculations. I'm not recommending Excel; I'm just noting that it's what everyone else uses. There are competing products like Openoffice.org's "Calc," which do pretty much the same things, but are free.

I used to prefer Lotus 123 back when it was a viable alternative but, as is so often the case, the most commonly used product drives out the competition, if only so that engineers have a common file format to collaborate with. Lotus 123 only went off sale in 2013, but it became a hassle to use it from a collaborative point of view around the year 2000.

The Excel product as it comes out of the box is pretty powerful and versatile, but it has a more powerful tool still contained within it (Visual Basic)—see the next section.

Many of the software tools used and taught in academia have no functionality for professional engineers greater than that offered by Excel, and are not as transparent as Excel.

It may be a pain to grind through someone else's spreadsheet checking that all the calculations are set up right, but at least you can, and you are certain that the program itself will do what is asked of it.

Naming Excel tabs appropriately facilitates the checking of calculations, and is practiced by most engineers, but many are still unaware of a feature which helps even more: naming cells. This makes self-checking (and more important still checking by others) a great deal easier.

To the left of the formula bar above the worksheet is the name box. You can use this to give individual cells or cell ranges a name (Figure 4.1).

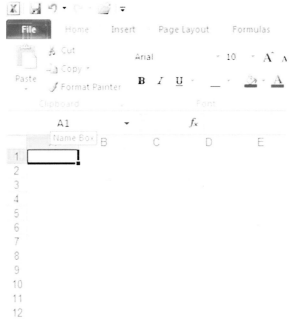

Figure 4.1 Screenshot of a blank MS Excel spreadsheet.

If you do this the formulae in your spreadsheet will have a format which allows you to check that they are correct without having to roam all over a multipage spreadsheet to see if the right box has been referenced.

To give a simple example, labeling cells C8–C10 with a description of their contents:

$$\text{Volume} = \text{height} \times \text{PI}(\cdot) \times (\text{diameter}/2)^2$$

Is a lot more transparent than:

$$\text{C10} = \text{C8} \times \text{PI}(\cdot) \times (\text{C9}/2)^2$$

For similar reasons, I avoid simplifying the equations I put into Excel. I create the terms in the equations within discrete brackets from obvious sources. This looks pretty ugly and unwieldy, but we should always think of someone else having to check your calculations, as well as perhaps having to come back to them in 15 years' time yourself.

Despite these simplifying measures, there will still be implicit assumptions or a required way of using spreadsheets which others might not know about, and we might forget over time. We should annotate the spreadsheet using comment boxes with this information, at its point of use. Separate documentation or instructions for use are much more likely to be lost over time than embedded versions.

Microsoft Visual Basic

Price: free with MS Excel.

Back when I started with computers, if you wanted them to do anything, you had to write the program yourself. Most of us began with a language called Basic, a cut-down version of the venerable Fortran.

Microsoft's version was called GW Basic (GW was commonly thought to stand for "Gee-Whizz!"—they are Americans after all). GW Basic's modern descendant is MS's Visual Basic (VB).

It allows us to automate spreadsheet functions, and write programs to do things which standard Excel cannot. However, this means that we are doing programming, in other words doing "a nontrivial job. . . requiring special skills and experience."

Those without these skills and experience need to consider the two main aspects of checking computer modeling programs: *verification* in which we check that all the elements are correctly coded, and *validation* in which we check that the model matches reality.

Past a certain (quite low) degree of complexity, computer programs do things we didn't expect them to. The best fully verified and validated commercial programs are thought to contain around 4% undiscovered errors.

Our own programs should be assumed to be far faultier. Writing your own program is rarely going to be the quickest way to solve a practical engineering problem, when the necessary validation time is considered.

That said, Douglas Erwin has written a useful book on how to use VB to assist with carrying out real design tasks (albeit only in industries based on organic chemicals)—see Further Reading at the end of the chapter for details.

If you have time for proper verification, writing programs in VB becomes a viable approach. This is usually only the case if the program is to be used many times. For example, I used some VB in writing the Excel spreadsheets I use for hydraulic calculations (of which there are dozens on every plant design I carry out), so it was worth paying to have the spreadsheets third-party beta-test verified.

I have used these programs with confidence hundreds of times since they were verified, even though the initial effort of producing them and having them checked was far more than I would have expended on manual calculations for a single project.

One-off programs are unlikely to be economic to properly test and verify, as Erwin acknowledges when he tells the reader that the VB programs which come with his book are not tested, and are consequently likely to generate error messages which he challenges users to fix for themselves.

Other numerical analysis software
MathWorks MATLAB
Single user license—depends on options, but certainly £10K + .

Matlab is commonly used and taught in academia, though hardly anyone uses it in professional chemical engineering practice. It underpins Simulink, a program which allows you to write your own dynamic modeling software (just like the IChemE tells us not to).

The IChemE's guidance on use of computers tells us why practitioners should not write their own models—unverified programs are error prone, and unvalidated programs are misleading.

It will undoubtedly be quicker to use a commercial program than to write, verify, and validate our own from scratch, but of course modeling is not design in any case.

As programs, Matlab and Simulink both may be fine when used as intended, but they just do sums. They do not do engineering.

PTC Mathcad
Single user license around £1,000.

Mentioned only for completeness, Mathcad mainly just does algebra. I have never seen it used in professional practice, probably because engineers can do their own algebra. It integrates with Creo, a drafting program from the same vendor, but process plant designers don't use Creo either.

Simulation programs
The dividing line between use and misuse of simulation and modeling programs is whether the IChemE CAPE guidelines are followed, and particularly whether model verification and validation is undertaken.

If you have a great deal of applicable data on the exact plant you are designing, and are designing many similar plants, you can go to the effort required to fit a modeling program to your plant, and write accurate models of your unit operations. Plant design then becomes a question of linking these blocks into an integrated model, and optimization of the model can be a valid proxy for optimizing the plant.

If, however, you are doing a one-off design, you do not usually have a great deal of information about the plant which will be built. Rather than using a validated model, you will be using the straight-out-of-the-box generic data and models, and optimization of this unvalidated model makes no sense. The errors in the model are very likely to be greater than the resolution of the optimization procedure.

Plant operators are the ones holding the information necessary to validate and tune modeling software. The contracting companies who design the majority of process plants do not have this information, and consequently make less use of modeling.

This is presumably why the most commonly used modeling software is written to support the oil and gas companies who are best placed to put in the data, time, and effort needed to validate modeling software.

There are quite a number of modeling programs, so I shall restrict myself to the most popular ones in the discussion to follow.

Aspentech Hysys, etc.

Single user license £10—20K.

Hysys is clearly written for process optimization in the oil and gas/petrochemicals industry, though it is nowadays commonly misused as if it were a design tool for all sectors in academia. Many green graduates I meet cannot attempt to "design" a plant without it.

If you google "Hysys," you will find Aspentech's site (where it is described in terms which match my understanding of its purpose), and a great many other sites by academics, where it is described as a design program of use across all process sectors.

Invensys SimSci Pro/II

Single user license £15K.

A steady state process simulator which is, I am told, perhaps more popular with contracting and operating companies in the oil and gas industry than Hysys.

The manufacturer's website states that "PRO/II offers a wide variety of thermodynamic models to virtually every industry," but its available unit operations are mostly limited to those of the oil and gas industry. As they themselves state:

Spanning oil & gas separation to reactive distillation, PRO/II offers the chemical, petroleum, natural gas, solids processing and polymer industries the most comprehensive process simulation solution available today.

Pro/II is, like Hysys, used as an optimization and debottlenecking tool in that industry, and it is similarly misused in academia to attempt to replace proper process design.

Chemstations CHEMCAD

Single user license £9K.

This was the simulation program I learned to use in university back in 1990. I never used it in practice, and have never seen it used in professional practice or academia, although the supplier's website makes it clear that it is still in production.

It looks to do the same things as Hysys, etc., and I have no reason to believe it does them any better or worse, other than its seeming lack of use in practice.

COMSOL Multiphysics, etc.

Single user license around £15,000.

COMSOL Multiphysics does a different kind of thing to the preceding products. It does not attempt to model whole plants, but does more detailed modeling of smaller systems. It is described by its manufacturer thus:

> You can model and simulate any physics-based system using COMSOL®. COMSOL Multiphysics® includes the COMSOL Desktop® graphical user interface (GUI) and a set of predefined user interfaces with associated modeling tools, referred to as physics interfaces, for modeling common applications. A suite of add-on products expands this multiphysics simulation platform for modeling specific application areas as well as interfacing to third-party software and their capabilities.

The add-ons which are most likely to interest chemical engineers are:
- Chemical Reaction Engineering Module.
- Heat Transfer Module.
- Computer Fluid Dynamics (CFD) Module.
- Mixer Module.
- Pipe Flow Module.
- Molecular Flow Module.
- Optimization Module.

It is claimed to be able to interface seamlessly with Excel, Matlab, AutoCAD, and Pro/II via other add-ons.

The problem? There is a great deal more going on in a process plant than physics, however "multi" that might be. It is not the software vendor's fault that someone might think that optimizing a highly simplified model of a subsection of a plant is optimizing the plant, but this is how it is misused in academia.

Other

There are also proprietary programs written as one-off specialist products which allow specific issues to be analyzed. They are designed to be more accurate than generic products, and I am told that, even when stripped of all components not related to the limited duty they are designed to, they may take weeks of processing time on a modern desktop PC to reach a solution.

Project management/programming tools

Plant designers need to be able to analyze and communicate the coordinated tasks which will be required to design, procure, construct, and commission the plant if they are to do accurate pricing calculations.

They will usually use the same tools for this as project managers will later use to keep the project on track and on budget, so the programs are usually a little overpowered for the use which plant designers will put them to.

Microsoft Project

Single user license £550.

The MS program is fairly commonly used by plant designers. Although other, more powerful specialist programs are probably more common tools for specialist project programs, everyone's familiarity with MS products makes "Project" easy to pick up.

Microsoft Excel

Single user license £110.

You can produce rough project programs in Excel, and sometimes it is expedient to do so as not everyone has Project, given that it is not included in all versions of the MS Office suite. Programs produced in Excel are not, however, really up to a professional standard of presentation, so it is preferable to use MS Project as a minimum standard.

Of course, you can export your MS Excel program into MS Project if you have formatted it correctly, as well as exporting MS Project data to MS Excel.

AMS Realtime

Single user license around £1,000.

I only mention this because it is the direct descendant of the program I learned to write project programs with, Artemis Schedule Publisher. Unlike Schedule Publisher, I have never seen Realtime used by a plant designer.

Oracle Primavera

Single user license around £3K+.

A far more sophisticated program than MS Project, which seems to have super-seded Schedule Publisher in the sort of companies I work for. As the promotional literature states:

> Primavera P6 Professional Project Management, the recognized standard for high-performance project management software, is designed to handle large-scale, highly sophisticated and multifaceted projects. It can be used to organize projects up to 100,000 activities, and it pro-vides unlimited resources and an unlimited number of target plans.

Not really for use by the plant designer, it is more for project managers, usually via a specialist project programmer, but if working in a company with such a specialty program, designers may use the program by proxy.

Computer-Aided Design (CAD): drawing/drafting

Autodesk AutoCAD/Inventor, etc.

Single seat license for basic package around £5K.

As with MS Excel, it doesn't matter whether AutoCAD is the best program—it's the one everyone uses, and all serious competitors make sure that their programs can export to Autodesk's dxf (Drawing Exchange File) format.

AutoCAD used to be a bit hard to learn because, for a long time, they resisted the now-standard (from MS products) meanings of mouse-clicks, return key and so on, but now the program allows you to use these as well as supporting the old-school "draffies" who still use it as if it were running under MS-DOS.

AutoCAD comes in a number of specialist flavors, with preinstalled content and other customizations suitable for various engineering disciplines, as well as a version specially for drafting P&IDs to US standards, though a company called Excitech will give you free files (see www.excitech.co.uk) to make "AutoCAD P&ID" compliant with British Standards.

Bentley Systems Microstation
Single seat license for basic package around £3K.

Whether Microstation or AutoCAD is the better product isn't the question— AutoCAD dominates the market. In my opinion, Microstation does itself no favors by being consciously so different from AutoCAD, such that there is a steep learning curve to master the most basic functions of what will probably always be the second-banana product.

Other than that Microstation is a perfectly good program, whose main advantages are keeping AutoCAD on its toes, and listening to its users. Virtually no one uses it for process plant design. Microstation has built-in simulation and modeling capabilities while AutoCAD does not, for people who like that sort of thing. (You know, theorists.)

PTC Creo
Single seat license for basic package around £4K.

Formerly known as Pro/Engineer, this is not widely used in process engineering. It is far more popular with those who do 3D drawings such as product and mechanical designers and architects.

Computer-Aided Design (CAD): process design
Bentley Systems Axsys.Process/PlantWise
Single seat license for basic package around £12K.

The suppliers say that it has been designed to allow rapid Front End Engineering Design (FEED) studies, and they describe it as follows:

> Interfaces with all the major process simulators, including HYSYS, AspenPlus, UniSim, and Pro/II, so you can use your system of choice and properly manage the resulting data. Data from multiple simulation runs can be easily compared and new design cases quickly generated
>
> Automatic creation of Process Flow Diagrams (PFDs) and Piping & Instrumentation Diagrams (P&IDs) using project-specific symbols and drawings that can be output to multiple drawing formats
>
> Integration with the major heat exchanger applications such as HTRI and HTFS for faster heat exchanger design

Use of Microsoft Excel to provide easy data entry and generation of intelligent datasheets and reports in your specific formats

A managed workflow that incorporates graphics, data, and work management for each user and tracks changes during the project, allowing users to revert to previous designs

Intelligent documents and data that can be transferred into detailed design products, including Bentley AutoPLANT and Bentley PlantSpace

Improved Front-End Engineering Design, which helps reduce capital expenditure as proven by users.

It also allows automatic pipe routing. This sounds like a great product, so I'm not sure why I had never heard of it before I researched this book.

Having an integrated suite of programs which combine drafting and design with modeling and simulation outputs sounds like an excellent idea on paper (especially for those following the academic modeling-as-design approach), but I wonder if each of the bits does its job as well as a dedicated program, or a professional engineer.

Computer-Aided Design (CAD): hydraulic design

Computational fluid dynamics (CFD) allows the visualization of complex patterns of fluid flow in a physics-based model. The output is often pretty, but does not necessarily match reality that well.

CFD isn't much to do with whole-plant design in any case, but I'll mention them in passing for those who think it is.

There may be occasions where CFD might come in handy to design a particularly challenging element of the plant but you'd be wiser to buy the service in from a specialist, than learn on the job.

COMSOL Multiphysics

This is one of the functionalities of the COMSOL product mentioned in an earlier section.

Matlab/Simulink

You can write your own CFD program using Matlab. You probably shouldn't though.

Autodesk® Simulation CFD

This dedicated program seems to offer some of the same fluid flow and heat transfer analysis features as COMSOL Multiphysics and Matlab. Being an Autodesk product, it presumably integrates well with AutoCAD.

Other

Microsoft Visio

Single user license £300.

Visio is flowchart software which is not intended for producing professional engineering drawings, but, like all MS products, it is easy to pick up, so it has become commonly used in academia as a substitute for the professional drawing programs most academics can't use.

It does have a good-quality "export to dxf" feature which allows its product to be brought in to AutoCAD for professionalizing, so all is not lost if beginning designers can only use Visio, but I personally don't teach it at all, as it is much simpler to go straight to the far superior AutoCAD.

I have seen it used by professional engineers in some interesting ways, dependent on its ability to link to Excel, for example producing dynamic hydraulic profiles which alter the relative position of drawing elements in response to underlying linked hydraulic calculations.

Microsoft Access
Single user license £110.

Microsoft's Database software, MS Access can be useful for document control.

FURTHER READING

Erwin, D.L., 2014. Industrial Chemical Process Design. McGraw Hill, New York, NY.

IChemE CAPE Working Group, 1999. The Use of Computers by Chemical Engineers: Guidelines for Practicing Engineers, Engineering Management, Software Developers and Teachers of Chemical Engineering in the Use of Computer Software in the Design of Process Plant. Institution of Chemical Engineers, London.

CHAPTER 5

The Future of Process Plant Design

PROCESS PORN

In my intercourse with mankind, I have always found those who would thrust theory into practical matters to be, at bottom, men of no judgment and pure quacks.

John Smeaton

Some areas of physics have, to outsiders, clearly lost themselves in abstraction. Seduced by the beauty of higher mathematics, they pursue things which seem to look right, even though they are piling unsupported speculation on top of itself many times over to get there.

The theorists responsible are now crying to be freed from the requirement to prove any part of their theories empirically. They think that they should instead be allowed to pursue mathematics and philosophy where they think they lead.

The partial differential equation entered theoretical physics as a handmaid, but has gradually become mistress.

Einstein

The problem is that mathematics and philosophy deal with what is plausible within their conventions, rather than the truth. Without a grounding in empiricism, physics of this type is pure self-indulgence, which is why the product of this lost school of physics is sometimes called "physics porn."

The philosophical tool which protects us against losing ourselves in abstraction in this way is the beefed up version of Occam's Razor known colorfully as "Newton's Flaming Laser Sword": "what cannot be settled by experiment is not worth debating."

We seem to have developed a similar problem in process plant design. In a computer model's mathematical space, many things seem plausible, but we only find out what is possible when we build the plant. Some are even generating things they call rules of thumb for design by repeated simulation, as if we had proven that such models are reliable analogues of the real world.

This approach is similar enough to physics porn in its lack of empirical support that we might call it "process porn." In the resultant academic discourse on process design, it seems now to be considered axiomatic that the approach followed by all professional engineers is hopelessly obsolete.

An approach based on higher mathematics, theoretical sciences, modeling and simulation programs, and network analysis techniques is now thought in academia to be the future of process design. This may have something to do with the fact that these disciplines and tools are those which academics know and, further, that the vast majority of chemical engineering lecturers worldwide have never designed a real process plant.

The idea that engineering is just applied mathematics and science is commonplace in these circles, but this is an idea held only by those who have never practiced, or have never reflected upon their practice.

All practitioners know that much of what they were taught in university is worthless in practice, and much that would have been of value was not taught (the supposed exception to this rule are French engineers, who according to the old engineering joke ask "So eet works in practice, but does eet work in theory?").

Universities feel that they have to staff their departments with scientific researchers, and few engineers want to do research—we want to be engineers, not scientists. Scientists are also a lot cheaper than chemical engineers. All of this was fine as long as staff knew that they were stand-ins for the engineers who were not available or affordable, but they have started making a virtue of their deficiencies.

As noted in a previous chapter, I hear that many research-led universities teach the following as a process plant design methodology:

- Look, in scientific literature, for bench-scale experiments which give possible process chemistry.
- Use these unproven techniques as the basis of a costing exercise which goes only as far as comparing feedstock and product prices.
- If the product sells for more than the feedstock, assume the process is economic.
- Use the unproven technique as the basis of a hysys model.
- "Optimize" the hysys model (which in this context means only getting recycles to converge).
- Grind through pinch analysis by explicitly defined rote methodology.
- Produce a short Word document which describes how they navigated the decision tree provided by the lecturer.

At the end of this time-consuming exercise, all we have is a worthless hysys model based on bench-scale experiments without any scale-up consideration, limited in scope to a reactor and an associated separation process.

We have given no thought to whether the standard hysys data and assumptions are valid in our case (they will probably not be), and we apparently think that getting recycles to converge in a computer model can be called optimization.

We have not required students to give any thought to cost, safety, or robustness, produced no engineering deliverables, and they have at no point been required to exercise the slightest judgment, imagination, or intelligence.

This state of affairs is not just useless, as it teaches the opposite of professional practice:

- Engineers don't ever use lab research as the basis of full-scale plant design.
- Engineers don't ever use modeling in place of design.
- Engineers don't ever ignore cost considerations.

- Engineers don't ever ignore safety considerations.
- Engineers don't ever produce "designs" that no one (including them) really understands.
- Engineers design by producing recognizable engineering deliverables.

In direct contrast, this academic approach represents, as I understand it, best practice according to researchers in "process design."

It is ultimately based upon approaches originally intended to allow beginners with no experienced supervisor to attempt a certain rather unrealistic and outdated kind of process design; approaches which have been stretched far beyond their originally intended purpose.

Some parts of the approach do have limited use, mostly in optimizing existing plant in certain industries, but these too are used out of context and without validation in the real world.

So let's unpick some of the ideas underlying the academic approach, and then consider some further questions about the future of plant design, bearing in mind how much of the core of process plant design is the same today as it has always been.

WILL FIRST PRINCIPLES DESIGN REPLACE HEURISTIC DESIGN IN FUTURE?

In a word, no. All is and always will be heuristic in the foreseeable future of engineering design.

We know this for sure because heuristic design is enshrined in law worldwide, and with good reason. Codes of practice and national and international design standards require heuristic design calculations to be carried out for safety purposes.

More theoretically, despite the hubris of some scientists, science has simply not advanced to the level where it can describe in sufficient resolution all the complexities of a proposed process plant.

Companies also have their own design manuals which require heuristic design checks to ensure that designs include the company's know-how.

Some companies now use tailored simulation program blocks representing the unit operations they most frequently design to carry out this duty, but they build into these blocks in-house empirical information obtained from past designs. This is not first principles design; it is empirically validated heuristic design. The program is just a container and vector for the empirical data, and its output is checked for sensibleness using further heuristics.

A process plant is too complex to be sufficiently fully described by any first principles model simpler than the plant itself, and any sufficiently complex model would be too complex to understand.

The purpose of heuristic design is to produce a good-enough model of the plant encompassing the state of the art which the process designer fully understands. A model which is slightly more accurate than the good-enough one is actually worse than one which is slightly less accurate, because it will also be slightly less well understood.

WILL PROCESS DESIGN BECOME A FORM OF APPLIED MATHEMATICS IN FUTURE?

This is the process pornographer's ideal, but it's just not going to happen, for the reasons given in the last chapter.

Process plant design will no more become applied mathematics than medicine will, as any practitioner understands, because:

As far as the laws of mathematics refer to reality, they are not certain; and as far as they are certain, they do not refer to reality.

Einstein

WILL PRIMARY RESEARCH BECOME THE BASIS OF ENGINEERING DESIGN IN FUTURE?

The academic papers freely available within academia are usually too expensive to access for practitioners outside the paywall, but even if we had free access to the papers, we understand how many high hurdles there are between bench and plant.

Academics frequently mark student design based on bench-scale academic research very highly, as it satisfies their desire for radical novelty, but engineers know that few things scale up well from bench to plant.

We do not give high marks for novelty—we give high marks for a working, safe, and cost-effective plant. We give low marks (i.e., fire you) for designing a novel but unsafe/unreliable/loss-making plant. The judiciary may also give you low marks in court.

So: No! Engineers are not going to start spending millions of dollars to build a plant which scales a process up by a factor of 100 from a bench-scale experiment any time soon.

WILL "CHEMICAL PROCESS DESIGN" REPLACE PROCESS PLANT DESIGN IN FUTURE?

The future of process plant design is envisaged by the IChemE in "Chemical Engineering Matters" as being to do with providing for human needs—food, water, medicine, and energy.

Douglas's original "Chemical Process Design" is based on a set of assumptions which do not hold in many of these sectors, and answers different questions to those asked of process designers in these areas. The school of Chemical Process Design developed from Douglas's ideas by other academics is less useful still in these sectors.

Some aspects of this approach are, however, finding favor in the petrochemical industry which matches most closely the basis of many of its embedded assumptions. This need not trouble us much in the developed world, as it is uneconomic to build new plants in this sector at our wage rates. The main use for its component techniques and tools is still the optimization, debottlenecking, and minor modifications to existing plant, *even in that sector.*

WILL NETWORK ANALYSIS FORM THE CORE OF DESIGN PRACTICE IN FUTURE?

Network analysis is all very clever, but it can only do one thing. Setting in stone the results of such an analysis as the foundation of design is very unlikely to be optimal.

1. The more integrated a plant is, the less controllable (and by implication safe) it is, and the harder it is to start the plant up. This costs the client money in commissioning and maintenance engineers' time, the provision of back-up equipment for start-up and so on.
2. In the specific case of heat integration, it should be noted that heat exchangers are not free. Putting in enough heat exchange capacity to approach the theoretical maximum possible heat recovery (as many academic network analyses suggest) will never be economically sensible. The interpretation of sustainability which is used to support such an approach runs counter to that in the IChemE's sustainability metrics.

This approach can have value in optimization of an existing plant, but in order to use it at conceptual design stage, we have to apply it to "optimize" an unvalidated model. Since optimizing such a model does not necessarily optimize the plant which is built, this is an example of mismatch between a design technique and the resolution of the model to which it is applied.

WILL PROCESS SIMULATION REPLACE THE DESIGN PROCESS IN FUTURE?

Rather than being the future, this is a very old idea, the fond hope of academics since computers were first devised.

Simulations are made of mathematics. Mathematics is perfect, but the real world is made of rather more complex and imperfect stuff, and contains even less perfect, even more complex people.

Materials and feedstocks are never perfect. Equipment is never perfect, and becomes increasingly imperfect over time. Operators are never perfect, and have a tendency to become increasingly imperfect over time unless well managed. Plants are never constructed exactly as per the original design.

I cannot find any research papers in which modeling and simulation programs are used in their straight-out-of-the-box format to design a plant, and the predictions of the model then validated against the real plant.

The most impressive support I can see for the validation of this approach is that after such programs are "calibrated" with large volumes of full- or pilot-scale plant data, they can predict performance of small sections of plant to a reasonable degree of accuracy.

This is, however, simply using the program to contain empirical data, with the program itself only filling in small gaps in the data, and even this approach has only "worked" with a couple of linked unit operations at a time, rather than a complete plant. A professional designer could probably have produced the design of a complete plant in far less time than it took for even this limited success.

So even if it became possible to accurately model the full complexity of the real world, it would take longer to program the model than to simply design the plant, and the part of the process design which a simulation describes is at best the mass and energy balance, and Process Flow Diagram (PFD).

Only someone who thought that safety, plant layout, hydraulic considerations, and cost were trivial side-issues would think it plausible that simulation could replace process design. Process design is in any case one small part of process plant design.

Even if this hurdle was overcome, plants designed by computer would be understood by no one. Plants understood by no one are not capable of verification as sufficiently safe, robust, and cost-effective. Following such an approach would be based on a misplaced faith rather than reasoned professional judgment.

Even if all of these issues were overcome, these programs do not produce the engineering deliverables which are the immediate point of design. They are not design tools.

WILL PROCESS PLANT DESIGN NEVER CHANGE?

Surely this has to be a "No" too, but some things have been conserved from the very beginning of chemical engineering.

If the defining concepts of Chemical Engineering, that is, of unit operations, mass, energy and momentum balancing, quantification and analysis, and so on, are lost, the discipline will no longer exist, but they have proved useful for some time now.

Similarly, as long as resources come at some cost, there will be selection pressure to maintain the same stages of design as are common to all engineering disciplines.

As long as the limits on our knowledge of physical sciences and computing power cause process system complexity to prevent us from making a completely accurate model of a proposed plant, so the limits of human brainpower will prevent the understanding of models beyond a certain level of complexity; yet professional responsibility requires us to understand what we are proposing: modeling cannot replace heuristic design.

But, in my professional lifetime, we have produced many useful new design tools and, to a far lesser degree, useful new design techniques. Things will undoubtedly change, but when I look back to what we thought the world would be like in 2014 back in the 1960s, I am hesitant to predict how they will change. I hope it will be as surprising as that was, though I'd gladly trade the internet for a personal jetpack.

I am, however, happy to predict that when professional engineers do change how they work, they will change to something which gives a provably safer, cheaper, or more robust product within the real constraints we have to work under.

FURTHER READING

Alder, M., 2004. Newton's Flaming Laser Sword, or: why mathematicians and scientists don't like philosophy but do it anyway. Philos. Now May/June (46).

Anon, 2013. Chemical Engineering Matters. Institution of Chemical Engineers, London.

PART 2

Professional Practice

CHAPTER 6

System Level Design

INTRODUCTION

The very essence of process plant design, the thing people employ chemical engineers to do, is system level design. By this I mean more or less the same thing as Pugh means by his term "Total Design," rather than the approaches called by similar sounding names in academia.

It is integrative, though I do not mean by this the academic "Process Integration," or "Process Intensification," I mean integration of the needs of all engineering design disciplines, and those who are to build and operate the plant.

It has a broad ranging vision—a good process plant design considers all the elements of design given in this book, in order to produce an appropriately detailed set of documentation to allow decision making in early stages, and construction in the last stage.

It is multidisciplinary, involving usually (as a minimum) civil and electrical engineers, as well as, to a lesser degree, construction stage staff, management, clients, and plant operators.

It is multidimensional, taking into consideration as an absolute minimum the cost, safety, and robustness implications of every decision.

It is iterative—a design evolves through successively better incarnations.

Lastly, the thing which makes it truly system level design is that process plant designers see in their mind's eye a complex system working as a whole. This is why all partial approaches entirely miss the point—it isn't about optimizing any one variable. It is about being able to imagine a completely integrated system which no-one can fully understand, but that the designer understands well enough to make it do what they want it to do in the way they say it is going to do it.

So how do we do this?

HOW TO PUT UNIT OPERATIONS TOGETHER

The main tools for system level design are the Piping and Instrumentation Diagram (P&ID), Process Flow Diagram (PFD), and General Arrangement (GA), which represent a great deal of our design deliberation in a concentrated form. We can see from them, at a glance, most of the things we need to consider in making the components of our plant work together.

We will use them in slightly different ways as design progresses, but the PFD encapsulates the integration of mass and energy balance, the P&ID system level control and integration, and the GA the physical and hydraulic constraints.

They are not just records of the designer's thinking, though this is an important function to allow the review of designs by others. Producing these documents forces the designer to consider the issues described in the last paragraph, and allows him or her to visualize the effects of their proposed solutions. They are therefore design tools as well as records.

None of the academic methodologies intended to serve the purpose of process integration are used in design practice because they are addressing a problem which professionals have already solved, in a more comprehensive and universally comprehensible way.

MATCHING DESIGN RIGOR WITH STAGE OF DESIGN

At the conceptual design stage, we have at least to get a broad idea of recycle ratios, as these can have a huge effect on main plant item sizing for certain types of unit operations, such as reactors and their often closely associated separation processes. This will involve generating a PFD and associated mass balance.

We need to get an idea of how physically large the plant is going to be, so that we can see if it is going to fit on the available site. We will need to produce a GA to do this, and carry out rough hydraulic calculations.

These hydraulic calculations will tell us whether we are going to have a completely pumped system, or make some use of gravity, a choice which will affect the plant layout and be seen on the GA. We will need to decide if we are going to use a batch or continuous process, a decision which will again affect layout and plant footprint.

The P&ID will be affected by these choices, and there are also choices to be made between software and hardware solutions to design problems, the solutions to which will appear even on early stage P&IDs.

At this stage we are probably designing unit operations using rules of thumb, without checking whether the units we are specifying are commercially available. We probably have fairly sketchy design data, have made quite a few simplifying assumptions, and have been given only a few resources to get to the desired endpoint.

Our aim is to see if we can fit the plant on our site, whether it is affordable and whether it is plausible from the point of view of cost, safety, and robustness. This is an initial rough screening, which the overwhelming majority of proposals fail. We cannot optimize such a design due to lack of definition, and we should not try, because we want to err on the side of caution.

If we are asked to produce a detailed design which we are willing to stand by as a fairly robust investigation, evaluation, and solution of the vast majority of design problems and tasks, then we need to look much harder. The same three drawings are, however, still going to be at the center of the exercise, but now they are primarily tools for collaboration with others.

System level design optimizes the whole-plant design, considering the implications of design decisions for our civil and electrical partners, installation and commissioning engineers, and plant operation and maintenance staff. We can send drawings (especially the GA which almost everyone can understand) back and forth with our design collaborators, installation contractors, and so on to take on board their opinions.

We can use the drawings to conduct design reviews with all interested parties. This is not to imply committees produce good designs, they do not. A plant "improved" by including layers of afterthoughts, or features which allow people to feel they contributed, is very likely to be a suboptimal design.

The process plant designer needs a strong vision and a willingness to challenge any suggested design modifications, inviting anyone making such suggestions to prove that they are improvements from the point of view of whole-plant cost, safety, and robustness. A plant "improved" in a way which, for example, maximizes profit or minimizes risk for the civil or electrical partner alone is unlikely to be optimal.

IMPLICATIONS FOR COST

There are many ways to consider the capital and operating cost implications of designs. Least capital cost is probably the most popular method, despite its shortcomings, but there are also evaluations based on whole-life cost, total cost of ownership, net present value, and so on.

A well-integrated design considers cost in the way the designer is asked to consider cost by the client. We need to set aside our preferences and prejudices and give them what they want if it is possible to do so.

We can design a least cost plant which operates for the defect liability period plus one day, or a plant with an excellent whole-life cost. We can design a plant based on the technologies the client's staff are used to even though there have been better technologies for decades.

None of these are wrong approaches, if they are the preference of the people who are paying. We may attempt to persuade clients to use another evaluation basis, but if it is not in some way cheaper, we are unlikely to succeed.

Generally speaking, from a capex point of view:
- More robust plant and materials may cost more.
- More automated plants cost more.
- Add-on safety equipment (but not inherently safe plant) costs more.

From an opex point of view:
- More robust plant and materials generally cost less to operate.
- More automated plants cost more to maintain, but save on operator time.
- Add-on safety equipment (but not inherently safe plant) costs more.

Note that from the point of view of both capex and opex, simplicity saves money.

IMPLICATIONS FOR SAFETY

We tend to consider safety issues based on legal responsibilities and professional codes of practice rather than client preferences in the first instance.

As discussed in the last section, afterthought safety (or "tinsel safety," as I used to call it back when my students used to "complete" a plant design with no consideration of safety and then decorate it with pretty safety features) costs money (it also reduces robustness), but system level thinking eliminates risk using techniques such as inherent safety.

The system level approach might actually save more money by eliminating layers of add-on safety features than the cost of the required safety measures controlling the risks which remain after minimization, substitution, and moderation of risks inherent in the design, and limiting the effects of adverse events.

Making the design error tolerant and simple may also be included under the heading of inherent safety, but these to me are basic design principles with effects beyond the realm of safety.

Any optimization which is carried out at conceptual design stage needs to address these paramount issues.

IMPLICATIONS FOR ROBUSTNESS

I was originally going to say here that robustness costs money, which is I think what most practicing engineers would say without a break for reflection, and is often true at the level of components, but is not necessarily so at the level of systems.

For example, which is better, a Daihatsu or a Bentley? Well, the Bentley looks pretty slick, but the Daihatsu has a reliability index of 40, and the Bentley 582 (Average reliability is 100, and the lower the number, the better—see "Further Reading" for source).

So robustness may be thought by many people to always cost money, but the robustness of simplicity, and well thought out integration of complex systems always saves money, and can trump the usually higher costs of robust components.

Clients may specify directly a period of operation of a plant, and/or they may give tender evaluation criteria which make clear whether they want a low capex, low opex, or low whole-life cost plant. Professional standards would prevent us offering a plant below certain standards, though these would be pretty low.

We could, if asked, design a plant intended as a display of wealth, showing that you just don't care how much it costs to run, because you can afford not to care. This is, however, much less popular in process plant design than in car design.

We can offer the same plant availability by using multiple cheap low-robustness units, or a smaller number of more robust items. The latter is often better from the point of view of whole-life cost, if not from a capex point of view.

RULE OF THUMB DESIGN

Rules of thumb are one of our main easy to learn tools in managing complexity. Real rules of thumb incorporate knowledge of the limits of operability, safety, and economics. They tell us where a working solution is likely to lie. They also tell us which approaches are likely to fail.

By "real" rules of thumb I do not mean those things with the same name now being generated using repeated computer modeling exercises. Real rules of thumb crystallize experience with real full-scale plant which is very similar indeed to the plant being designed. Computer modeling is theoretical first principles design unless the model has been validated against real full-scale plant. You cannot generate experience from theory, only from practice.

It should be noted that all rules of thumb are specific to a set of circumstances, and contain implied assumptions. They should not be applied outside these specific circumstances without the greatest of caution. Ideally, they should not be applied outside their specific case at all but, practically, we may have to bend the rules a bit on occasion. If we do this, we need to know that we are doing it, and reflect this in our degree of confidence in our answer. Complacency has led to disaster on many occasions in engineering.

FIRST PRINCIPLES DESIGN

This is not how professional process plant design is done, because anything designed from first principles is essentially a prototype, and its operators are test pilots.

If we are forced to carry out first principles design, there will undoubtedly be teething problems. Such a plant may not be safe, is unlikely to be robust, and will probably not be cost-effective.

So this is not a professional design approach, and there is a fair chance that its product cannot be made to work at all. This will come as a great disappointment to the client who spent large sums of their money on what they thought was a competently executed design.

This is not to say that there will not be occasions on which you are asked by a client to do something novel without being given the resources to characterize the problem well enough to avoid first principles design.

This happens to me fairly frequently, especially now that I have an academic title. It can lead to some difficult conversations. The following should be markers for possibly crossing a line in this area:
- Being asked to scale up a process by a factor of more than five;
- Being asked to design a unit operation (or worse still a whole plant) based on bench-scale tests of kit for which there are no full-scale references;
- Building the first plant of any type or at significantly increased scale.

I am not saying that you should never design a prototype (I have designed a few myself), but that you need to know that is what you are doing, and not confuse this R + D activity with normal professional process plant design.

DESIGN BY SIMULATION PROGRAM

Many simulation programs now come with rough costing data built in, but unless you add your own safety and robustness elements, they are not considered, and you can easily "design" plants you don't understand very well with no safety factor.

There is no need for a simulation program operator to keep the plant simple enough to understand it, or to have a model of any part of the system in their head. There is no need to consider the requirements of other disciplines or stakeholders, nor any tools to produce collaborative documents.

These factors should give us grounds for grave doubts about the use of such programs in anything other than the most tightly constrained circumstances. I simply cannot disrecommend strongly enough their use for "process optimization" at the conceptual design stage, ignoring most of the important factors in true system level plant optimization.

Simulation and modeling programs can, however, be of use for equipment suppliers to specify standard products, assemblies of standard products, or standard package plants. In this application, the default values in the program are replaced by users with real operating, thermodynamic, and costing data, so that the program is really operating as a convenient dynamic repository of empirical data.

There can still be some limited use of imagination and even a little innovation under these circumstances, but we are in my opinion shading into a modeling program operator/salesman role here.

SOURCES OF DESIGN DATA

Client documentation

Usually our client will give us some documented idea of what they want, and may well have gathered some data to assist us in our design. Clients, however, frequently attempt to disavow any responsibility for the information they supply, and try to pass on all design responsibility to design companies.

Designers will need to do what they can to exercise due diligence in checking that the client data is correct. There are potential opportunities here as well as problems. Sometimes the correct information or approaches can be more competitive than the suggested ones, or an alternative endpoint can be better, safer, or cheaper than the one the client asks for. Commercially minded engineers (which should be all of us) should be on the lookout for these ways to get ahead of less flexibly minded competitors.

Design manuals

It is common that companies will specialize in designing certain sorts of plants, and will have codified their know-how in in-house design manuals. These are often mandated as the source of information and approaches to be used by the company's designers.

Going off-piste is at the designer's own risk, and is frequently a sackable offence. It also exposes designers to the possibility of legal action in future.

That said, I have worked at places which had bad design manuals based on modeling program output, and felt a professional obligation to challenge the manual. This does not always go down well, but we are not automata, and "I was just following orders" hasn't historically been a good legal defense. Professional judgment is required of professionals.

Standards

Governments, national and international standard institutes, and trade bodies also codify know-how about what works and what does not (especially with respect to safety issues) in standards and codes of practice. Failure to adhere to codes and standards may be more serious than failure to follow design manuals. It may be illegal and even imprisonable.

National, regional, and international standards may well have conflicting requirements. The choice of which ones to follow is often dependent on the sector which the plant falls into, or conventions with respect to the type of equipment being specified. For example, the oil industry works largely to the American Society of Mechanical Engineers (ASME) and American Petroleum Institute (API) standards, and air pollution control equipment suppliers may look to German TA Luft standards.

Make sure you are following the most up-to-date version of codes and standards, and don't follow codes and standards blindly. Exercise professional judgment. Partial compliance with a written standard for a good reason is better than slavish compliance where it is inappropriate.

Don't play mix and match to get round a tough standard, and have a consistent philosophy from the conceptual design stage on issues like the acceptable size of fugitive emissions. Make it a good one, as it is expensive to change it later. Consistency of philosophy makes for ease of comprehension by operators and other engineers later in the design process.

Note that in countries with well-developed regulatory environment, regulatory authorities may specify codes and standards to be followed, and offer their own statutory guidance. In less-regulated countries, the minimum standards consistent with professional ethics are applicable.

Manufacturer's catalogues and representatives

This can be a rich source of knowledge. In order to produce a design integrated at depth, you need to have a very deep knowledge of the parts of the thing you are designing (or know someone who does). Equipment salesmen are not merely an annoyance as some designers think. You can work with them to produce very well-integrated designs.

More experienced engineers

For philosophers the argument from authority is a logical fallacy, but I'm not going to worry about the opinion of people who doubt whether the sun is going to rise tomorrow.

For new engineers, discussing their ideas with someone who has the feel for design through long experience is far better input than the most rigorous mathematical modeling exercise.

We old guys might seem a bit negative to newcomers, as so much of experience is knowing what doesn't work, and the practical constraints on plant design, commissioning, and operation.

As neophytes ("noobs") become more experienced themselves, they come to see that the constraints are not stifling but are the rules without which the game would be no fun.

Pilot plant trials/operational data

Sometimes we have the luxury of information on the operation of an existing very similar plant at the proposed site, or at least a similar type of plant at a similar site.

Sometimes we may have data from a similarly sized pilot plant of the proposed type at the proposed site. Sometimes the data we have is for less similar plants.

We need to be cautious with such data, and to be honest with ourselves about how much trust to put in it. Scale effects need to be considered. A pilot plant less than say 20% of the rated capacity of our plant might not be that strong a guide as to the constraints on operation of a full-scale plant.

Other sites and other technologies may be less similar than we think to our proposed site and technology.

Such data may be used to validate a computer model. Where this is done well (as it is done, e.g., by Air Products) this is a great aid to the designer. A validated model is, however, only true for the plant whose data has been validated. It is no more capable of predicting the operation of a plant 10 times as large as a physical pilot plant would be—quite possibly less so.

Previous designs

The same companies which have enough experience to have good design manuals also probably have staff who have designed previous similar plants, and commissioning engineers who have commissioned such plants. These are the people who know what does and doesn't work.

There will in such companies be access to drawings and calculations used to design previous plants, costing data, and an opportunity to improve on the last design based on commissioning and operational experience.

That is why clients like to buy from companies with a track record, and new designers would be wise to try to work for such companies.

"T'Internet"

The internet looks quite different outside academia, as professional engineers rarely have free access to scientific papers. This is not, however, a big problem: most scientific papers are about experiments at too small a scale to be of much use to full-scale plant designers anyway.

A far more valuable use of the internet is the free 24/7 access to manufacturer's data, literature, and drawings. Nowadays we can choose unit operations based on detailed information, and insert an accurate AutoCAD version of the chosen kit into our drawings without even talking to manufacturers.

A word of caution—the internet is undiscriminating in its content. A feel for the reliability of internet sources has to be developed. If a page looks like it was produced by a barely literate teenager it is usually obvious that it is not a reliable source of information. Obvious puff pieces for equipment manufacturers abound in Wikipedia articles as well as on their own websites. The most useful questions to interrogate web pages with are "Says who?" and "Based on what?" If you cannot see solid backing for claims, best not rely upon them.

One more thing: I would advise you not to go on web forums asking more experienced engineers the most basic details of how to design the thing you are designing. They will wonder which Muppet asked you to design that thing when you have no clue where to start. This reflects badly on both you and your employer.

Libraries

Yes, actual physical libraries, with books made of paper—they still have them. Large public libraries may have handbooks and textbooks which are of use to the designer. It is also usually possible for alumni to get free library access at the university they attended. Sometimes it is cheaper to take a trip than to buy a copy of a technical resource, as these can be very expensive.

FURTHER READING

http://www.reliabilityindex.com/manufacturer.

CHAPTER 7

Professional Design Methodology

Aeroplanes are not designed by science, but by art in spite of some pretense and humbug to the contrary. I do not mean to suggest that engineering can do without science, on the contrary, it stands on scientific foundations, but there is a big gap between scientific research and the engineering product which has to be bridged by the art of the engineer.

British Engineer to the Royal Aeronautical Society

INTRODUCTION

Each of the stages of design produces increasingly precise deliverables. We start with broad sketches and rough back-of-the-envelope calculations and we get down to drawings accurate to the millimeter and calculations good enough to purchase expensive equipment which will reliably do a specific duty.

So each stage of the design process has a natural resolution, or granularity. It will be resource-intensive to produce a more precise definition of the design at any given stage than is conventional, and the additional precision will be of no benefit to those commissioning the exercise. It will be wasted effort.

One of the mistakes made by early-career designers is to attempt to design unit operations in great detail at the conceptual design stage. They often find that the information they need to do the calculation as they were taught at university (at best, as set out in Sinnot & Towler) is simply not available.

Both information and design challenges are generated during each stage of design. Much of the information needed to carry out detailed design of unit operations may never be publicly available, but specialist suppliers will usually have it, or an alternative design methodology which does not need it.

Newcomers to this approach, coming from the certainty of mathematics and pure sciences which is taught in many places as chemical engineering, may find it a little odd, even a bit flaky. Why not institute a scientific research program to resolve all the uncertainties and collect all the missing data? Why not produce a simulation based on this program so detailed that the design process is more or less just a process of interrogating the simulation?

Why not? Because this would cost a great deal more, take far longer, and produce less good results than just having professional engineers design and build the plant, and no one is going to pay you to do that. For example, a company I used to work for tells me that they are producing a reliable, empirically validated model of a single unit operation. After 2–3 years of development, it is almost (but not quite) as good as the 30-year-old rule of thumb based design approach which they are still using to actually design plants.

So if you are going to use modeling properly, your competitors can give the client the information he or she needs for a tiny fraction of the resources you propose to use, probably even as a "free" favor. Process plant design is a commercial activity. Any methodology which does not recognize that will remain at best an academic curiosity.

DESIGN METHODOLOGIES

There is really only one design methodology in all design activity, the iterative staged approach described in this book.

However, designers may use discipline-specific design techniques, design philosophies, and design support tools. In circles where no one designs plants for a living, all kinds of "design methodologies" can grow up.

Setting aside the academic "Chemical Process Design" methodology discussed previously, there is a handy list of examples of "Design Methodologies" in Koolen's book "Design of Simple and Robust Process Plants" as follows:

- Inherently Safe Design.
- Environmentally Sound Design.
- Minimization of Equipment.
- Design for Single Reliable Components.
- Optimize Design.
- Clever Process Integration.
- Minimize Human Intervention.
- Operational Optimization.
- Just in Time Production.
- Design for Total Quality Control.

He then proposes to roll all of these into one in order to obtain "an optimally designed safe and reliable plant operated hands off at the most economical conditions."

Koolen's background is plant operation, rather than design. In common with many academic approaches, it is clear from his description of the process that this is a technique for design of modifications to hydrocarbon processing plants with extensive operational data supporting a modeling exercise.

Many of the things he calls "design methodologies" are used in professional practice, but none of these are really design philosophies or methodologies, nor are they all universally applicable or without conflicting interactions.

To take one example, "Minimize Equipment" has the subcategory "Avoidance of more reactor trains by development of large reactor systems." Theorists, researchers, and idealists will always favor scale-up (making one big novel reactor) over scale-out (having lots of small proven reactors).

Professional designers will also consider the time and expense of the implied development program.

Scale-up needs to balance the turndown of one large reactor against two or three smaller reactors. You can achieve a larger turndown with two or three smaller reactors by turning some off. If the market for your product reduces, and the single larger reactor has insufficient turndown, the site which went for scale-up rather than scale-out may close.

Notice also the clash between this approach and Inherently Safe Design's desire for small reaction masses. There is simply no way to avoid the need to apply professional judgment.

I mention this book because I have far more sympathy for his approach than those proposed by academics without operational experience. None of his "design methodologies" are terrible ideas in all settings—they all have their place, though for some of them it is in process optimization rather than design.

It seems a clever approach to the problem it sets out to address, producing a hydrocarbon processing plant as simple to operate as a washing machine. I don't know if this problem has ever actually come up, let alone if anyone has actually applied the methodology to it, but it looks (to someone who has never been asked the question) as a decent place to start answering it.

There are approaches used in operational companies to carry out "designs" of sufficient quality to allow operational companies to supervise detailed designs done by contractors on their behalf. These are the foundation of the approach used in Koolen's book, as well as any industry input to academic design approaches. This is not, however, process plant design as I define it. This is optimization of an existing design by operating company staff.

I know that there are many theoretical approaches in academia and in operational companies for the thing they call plant design, but in professional process plant design practice there is one basic methodology, used worldwide and across sectors. In the remainder of the book, I will describe this approach in enough detail to hopefully allow a beginner to master the basics.

THE "IS" AND "OUGHT" OF PROCESS DESIGN

Process design deals with what ought to be. It is not a scientific description of something which already exists, but a practical creative activity aimed at bringing into being something thought desirable.

Much of academic discussion of process design is to do with how design ought to be done, rather than how it is done. Such discussion is normative, rather than descriptive. Without claiming that what is, is what ought to be, I will offer an approach based on a description of what I believe to be the modern consensus approach.

What process design ought to be is a way to imagine, select, evaluate, and define safe, cost-effective, and robust solutions to the problems inherent in the design problem we have been asked to address. All ways of achieving this are good. We are paid to be engineers, not ethicists, nor social reformers.

RIGHT VERSUS WRONG DESIGN

There is no entirely right design, but there is an infinite number of wrong ones. Being sufficiently right really matters if someone is going to spend millions of dollars on the thing you have designed.

Engineers tend to only design things for construction which they really understand. You might think you can grind through BS PD5500 or EN 13445 and design a pressure vessel as well as the next person, but if you tell a supplier to make such a vessel to your specification, you are taking responsibility for the integrity of an item made by others. This is why we tend to make those who supply equipment responsible for its design.

We need to be able to do enough design to give suppliers' proposals a quick check over for reasonableness, but they are far more likely than us to have the know-how to make a specialist item in a safe and cost-effective way.

INTERESTING VERSUS BORING DESIGN

A good scientist is a person with original ideas. A good engineer is a person who makes a design that works with as few original ideas as possible.

Freeman Dyson

You need a damn good reason to be interesting as an engineering designer. "Interesting" or revolutionary design is nowhere near as likely to work as boring old "normal" design, where there are only incremental changes in design approaches which are known to work.

Academia is very keen on interesting design, as the research which they are familiar with is judged to a large extent on its interestingness. Practitioners need something which will definitely work, rather than an interesting research project.

Vincenti gives a number of examples of this phenomenon in his book "What Engineers Know and How They Know it," most notably the case of the development of the jet engine in Germany and the United Kingdom.

After early success with centrifugal engines, the Germans designed an axial flow engine years ahead of its time, very similar to modern jet engines. There were few suitable materials to make such an engine at the time, even if they had been working under peacetime conditions. The German Junkers engine consequently had a life in service of around 12 h, before it needed replacing with a new engine. It was a terrible waste of resources. Cost, safety, and robustness had been ignored in favor of novelty and elegance (Figure 7.1).

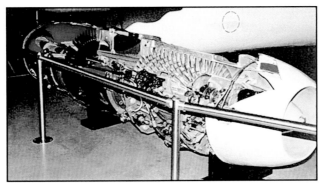

Figure 7.1 The Junkers Jumo 004 engine. *Image reproduced courtesy of National Museum of the US Air Force.*

The British jet engine designed at the same time was ugly, but it was designed to be produced under the prevailing circumstances. It was a "boring" design, but it had a very long service life, and went on to set world speed records (Figure 7.2).

Figure 7.2 The Whittle W2-700 engine.

As with all in engineering, there is a balance to be struck. Dyson says "as few original ideas as possible." He doesn't say "no original ideas."

CONTINUOUS VERSUS BATCH DESIGN

There is often no real choice if you want to be both right and boring—traditionally much process design was batch, but equipment tends to be better utilized in a continuous design.

Batch processing is still, however, very common in low-volume/high-value applications. I have made this section unusually long (with kind assistance from Keith Plumb from the IChemE's Pharma subject group), as it is an often neglected area.

Although continuous processing is used to make the high tonnage materials produced by the oil and gas, bulk chemical, and water sectors, a far greater number of materials are produced in batch processes than in continuous ones. In a recent survey carried out in a food ingredient factory, the company was using over five thousand different raw materials. All the small chemical plants using batch processes outnumber the total number of plants running continuous process many times over.

Specialty chemicals, pharmaceuticals, cosmetics, and food are nearly always made batchwise. Around 25% of chemical engineers work in these sectors, so batch processing is important to chemical engineers. For some process sectors it is critically important.

It is not easy to say why batch processing seems to be the poor relation of continuous processes with respect to publications but this is undoubtedly the case. A quick scan of the index of Sinnot and Towler or Perry's Handbook shows how little space is given to this topic.

Sinnot and Towler have three entries in the index, two paragraphs on batch distillation, and nothing on batch heat transfer. Perry does a little better with 14 entries in the index and several pages on some topics. However, this amounts to considerably less than 1% of the content of the handbook (as does this section of my book!).

Why use batch processing?
Apparent simplicity
On the face of it, batch processes appear to be simply a scaled up version of the process that a chemist would use on a laboratory bench.

This makes batch processes attractive if you do not want to spend too much time and money on development work. It is easy to get from bench to commercial scale cheaply and quickly if all you are doing is making the kit bigger.

For some sectors, such as specialty chemicals and food processing, the huge number of products and, in some cases, the short product life cycle means that getting to commercial scale cheaply and quickly is very important.

For pharmaceuticals, getting a product on the market quickly is important because of the limited patent life. In general at least half of the patent life is lost during clinical trials and process scale-up. Batch processes that can be scaled up quickly are at a distinct advantage.

However, once you look at a batch process in detail, it soon becomes clear that in practice it is not as simple as it first appeared. The chemistry is frequently poorly understood and the nonsteady state regime of batch processes makes them difficult to model.

Batch processes are therefore quick to scale up but difficult to optimize and their efficiency is consequently usually far below that of continuous processes.

Flexibility

Flexibility is a great advantage of batch processes. If you have a generic set of batch processing equipment then it is often possible to make a wide range of products. Some batch plants make hundreds of different products using similar processing methods and plant.

Multipurpose plants can be designed for a generic group of products and are frequently designed without any knowledge of the actual products to be made.

This is achieved by using a facility equipped to work with temperatures in the range $-100°C$ to $250°C$, pressures from high vacuum to 6 bar g and pH from 1 to 14.

Such plants are often made from highly corrosion resistant materials such as glass, graphite and tantalum, to allow them to handle a wide range of chemicals, unknown at the design stage.

Solids handling

Many products in specialty chemicals, pharmaceuticals, food, and cosmetics are solids or semisolids. These products can be difficult and expensive to handle safely and economically in continuous processes, particularly at the small scale.

In the case of pharmaceuticals, even the smallest scale commercially available continuous solids handling equipment may be of the order of 10 times larger than required. Small-scale batch solids handling equipment is generally much cheaper, less complex, and easier to maintain than continuous equipment.

Batch integrity

One advantage of batch processing is that it is possible to identify when the processing of a given sample started and stopped. This means that if the material manufactured does not meet the specifications, it is possible to reject just the particular batch that failed without needing to reject other material.

This used to be a highly important part of quality assurance when we depended on end-of-batch analysis and testing for quality assurance. As more online analysis becomes available, batch integrity is becoming a less important part of quality assurance and many products are released based purely on the online analysis.

However, there remains a business risk in relying solely on online analysis and many companies still like to retain batch integrity to minimize their exposure to the consequences of release to market of out-of-specification material.

The main batch design requirements

There are two major differences between continuous and batch processes; the non-steady state nature of unit operations and the importance of time-related sequences of operations.

One of the major differences is that it is not possible to summarize the details of a process using a Piping and Instrumentation Diagram (P&ID), as you would with a continuous process. It is necessary to have other documents to indicate how the process changes with time.

Nonsteady state

The nonsteady state nature of batch processing impacts on all unit operations. To illustrate the point, the three most important aspects are examined in the next section.

Batch heat transfer

If you heat or cool a batch of liquid in a vessel, the temperature difference between the heat transfer fluid and the batch of material changes with time and so does the outlet temperature of the heat transfer fluid.

If you have the simplified case of the cooling a homogeneous batch of material with an internal cooling coil then the heat transfer equation becomes:

$$\ln\left(\frac{T_1 - t_1}{T_2 - t_1}\right) = \frac{WC}{Mc}\left(\frac{K - 1}{K}\right)\theta$$

where:

T_1 = initial batch temperature
T_2 = final batch temperature
t_1 = cooling fluid inlet temperature
W = mass flow of cooling media
C = specific heat of cooling media
M = mass of the batch in the vessel
c = specific heat of the batch
$K = e^{UA/WC}$

A = heat transfer area

θ = time.

Even for this relatively simply case, the equation is quite complex and, as can be seen, includes time. For the fairly common case of using an external heat exchanger and a liquid being fed into the vessel, the equation becomes very complex.

Batch distillation

In the case of batch distillation, the concentration of the liquid in the reboiler (the batch) will be changing with respect to time as the more volatile components are driven off. This means that the temperature in the reboiler will rise over time and the concentration of components in the fractionating column will change with time.

To maintain the required concentration at the top of the column, the reflux ratio has to be increased over time. A point will be reached where it is no longer possible to maintain the top concentration, and distillation will have to stop or the top product be diverted to a separate receiver.

Batch distillation can be used to produce multiple fractions, but instead of the flows being taken off at different points in the fractionating column, the fractions are determined by time.

Batch reaction

In a steady state continuous stirred tank reactor (CSTR) the conditions remain constant within the reactor, but in a batch reactor, concentration, temperature, pressure, viscosity, density, etc. can change with time. An agitator which was appropriate at the start of the process may be much less suitable at its end.

The contents of the vessel may be more heterogeneous than in a continuous process, and different parts of the batch will consequently see different reaction conditions during the period of the reaction. This is one of the reasons why apparently simple batch reactions are in fact very complex. This lack of homogeneity usually becomes more of a problem as the scale increases and it is increasingly likely that unexpected reactions occur that have a serious impact on product quality.

Batch sequencing

To be able to calculate the capacity of a batch plant it is necessary to consider the sequence of operations and whether these operations take place in series or in parallel. Most batch processing plants have a number of parallel streams of the same series of operations.

The plant capacity is usually based on marketing demand forecasts for the products that the plant is being designed to manufacture. Keith Plumb tells me that the only thing that you can know with absolute certainty about such forecasts is that they will be wrong.

The design tools for batch sequencing and capacity calculations are similar to those used for engineering project management, a combination of Gantt and PERT charts. However, instead of basing the charts on flows of resources, they are based on mass flows and document the mass and energy balance.

Energy balance and utility requirements

The energy balance will be based on the heating and cooling requirements for reactions, distillation, and other unit operations, as is the case for continuous plants. However, the process sequence will determine where and when energy needs to be input to or removed from the system.

The input and removal of energy will be time-dependent and nonsteady-state, which makes energy recovery difficult. Heat integration techniques are even less appropriate to batch process plant design than continuous process design.

Even working out accurate predictions of working utility requirements is impossible, as the system is too poorly characterized for these to have any certainty.

The simplest (and least wrong) approach is to calculate the maximum utility requirements for the worst-case scenario of process steps coinciding and then apply a "diversity factor," basically a guess of how much of the maximum possible load will occur in practice based on practical experience of batch processes.

This is not always done well, and some plants are consequently constrained by lack of utility supply, requiring additional capacity to be retrofitted. In other cases, too large a capacity with insufficient turndown is supplied, leading to controllability and efficiency problems requiring remedy.

Despite there being many remaining problems in the field of batch design, it is still very popular in certain sectors, and there are a multitude of successful batch processes in operation today.

SIMPLE/ROBUST VERSUS COMPLICATED/FRAGILE DESIGN

There are so many quotes about the importance of this that I am spoilt for choice, but how about starting with someone considered by many to have been the first engineer?

Simplicity is the ultimate sophistication.

Leonardo Da Vinci

Van Koolen says in "Design of Simple and Robust Process Plants": "*A process plant should meet the simplicity and robustness of a household refrigerator.*" This might be a bit over the top in most cases, but good designs are (in a phrase widely attributed to Einstein) as simple as possible but no simpler. This is not, as some think, quite the same thing as Occam's Razor—it contains an additional note of caution against oversimplification.

Complex designs shouldn't be ruled out, sometimes they are needed, but they should be evaluated bearing in mind the fact that simple plants are easier to understand, easier to analyze (as Popper points out), more robust, more likely to be getting to the root of design challenges rather than piling afterthoughts on top of each other. More operable, more maintainable, more commissionable, more reliable, more available, more robust. What's not to like?

A classicist turned computer scientist puts it well:

There are two ways of constructing a … design: One way is to make it so simple that there are obviously no deficiencies, and the other way is to make it so complicated that there are no obvious deficiencies. The first method is far more difficult.

C.A.R. Hoare

There is an unexamined axiom in the "simple and robust" approach similar to that in academic approaches, the optimization of a small number of variables. Do we really need all process plants to be operable by the general public? Or are we shooting for a simpler design than is necessary if we blindly apply this approach?

A typical engineer, Koolen proposes to quantify simplicity with a view to allowing it to be systematically reduced. Do we really need to quantify simplicity? I have designed a few small package process plants which are to be operated by the general public, and I did not apply Koolen's mathematical/theoretical approach. I knows simple when I sees it.

It seems that perfection is reached not when there is nothing left to add, but when there is nothing left to take away.

Antoine de Saint Exupéry

Lessons from the slide rule

Before computers or even electronic calculators, engineers had slide rules. They couldn't easily add and subtract and had to guess where the decimal point was. This meant that engineers needed to be quite adept at mental arithmetic, and only worked to three significant figures.

My students get nervous when I round things up, do rough sums in my head and so on, like engineers of my generation. They believe that all 10 of the figures on their calculator displays are significant, even when I have set them a problem with two significant figures in the question data.

Process plant design engineers are probably kidding themselves if they think that they are working beyond three significant figures. Their underlying data is probably at best to this degree of precision. The extra decimal places on calculator and computer screen are spurious precision.

A feel for the sensibleness and reliability of our numbers is what really matters. Our modern tools seem to be taking this judgment away from new engineers.

Perhaps universities need to start using the QAMA calculators you can get now that require you to estimate the answer before they will give it to you.

Estimation/feel

He … insists that no mathematical formula, however exact it may appear to be, can be of greater accuracy than the assumptions on which it is based, and he draws the conclusion that experience still remains the great teacher and final judge.

James Kip Finch

A feel for the potential error associated with your answer, and its consequent meaningful precision is very important in engineering (even those who are not Finch's assumed "he"s). It is related to the margin of safety and turndown required to make a plant which will work.

New Scientist magazine conveniently gave us a word for this discipline, "olfactorithmetic," or the ability to notice if a number "smells" wrong. It will come with practice, but you can start to get it by always remembering the compounded uncertainty of your sources of data, the imprecision of your heuristics, and the probability of error.

As a rule of thumb, remember that every engineering design method is based on assumptions and simplifications; your original design data has an associated degree of uncertainty, as does any chemical/physical data you are using. If you have not identified, verified, quantified, and multiplied these assumptions, simplifications and degrees of uncertainty, your calculations are at best very rough.

There is nothing wrong with rough calculations, as long as they are combined with professional judgment. If I have sized a piece of equipment using three independent rough sizing methods which have a decent track record of success in professional use, and used professional judgment to make sense of the answers, I am far happier that I have a robust solution than if I had commissioned a five-year bench-scale research study.

So, avoid spurious precision—be honest with yourself about how much you really know, and specify equipment with a margin of safety plus turndown.

SETTING THE DESIGN ENVELOPE

We might call much of what is taught in university "Training Wheels Design." Just as we teach kids to ride a bike by minimizing the complexity of the task with a couple of ancillary wheels, we minimize the complexity of the design task for beginners with an assumption of "steady state"—all parameters (flow, composition, temperature, pressure, etc.) are assumed to be constant during the life of the equipment.

A real plant operates for most of its life within bounds set by the effectiveness of its control system, a significant part of its life in commissioning/maintenance conditions outside these bounds, and has to be sufficiently safe when operating well out of bounds in an emergency situation.

Although a plant will spend most of its life operating within bounds (though not actually at steady state), the maintenance and emergency conditions we need to accommodate may well have a larger effect on the limits of plant design than the requirements of quasi-steady state normal operation.

So we have to design a plant which will handle normal variations for extended periods of time, maintenance conditions for shorter periods of time, and extreme conditions for short, but crucial periods.

Each parameter therefore has a range of values, rather than the single value it has in the steady state case. So real plant design may have dozens of parameters at each stage, each of which has a range of values. The range of values will be associated with a range of probabilities, similar to a confidence interval. If our design is good, extreme values will be experienced with a low probability, and average values will be commonplace.

We frequently use confidence intervals to decide on the upper and lower bounds of incoming and outgoing concentrations of key chemicals. Performance trials are frequently statistically based, so we are in effect already working to a confidence interval in our product specification. We do not, however, design to the specified confidence interval: many engineers usually go up at least one standard deviation, such that a 95% confidence interval specification from the client leads me to work to a 99% confidence interval design.

We then need to permutate these ranges of parameters to generate design cases, representing the best, average, and worst cases we can imagine across important permutations.

For example, if I am designing a sewage treatment works, it needs to work in the middle of the night when no one is flushing a toilet and most factories are closed, as well as at peak loadings. It needs to work during dry weather, and during a 100-year return period storm. It needs to work sufficiently well when crucial equipment has failed.

When it rains hard, we can initially get a sharp increase in solids and biological material coming in through sewer flushing, and afterwards we get large volumes of weak sewage, mostly rainwater.

So I need to design to a low probability/high strength/high flow scenario, a medium probability/high flow/low strength scenario, a medium probability/low flow/high strength scenario, and a high probability/medium flow/strength condition.

I can imagine those who think chemical engineering to be petrochemical engineering thinking this only applies to water and sewage treatment, but this is just a specific example of a general condition: oil is a natural product too, whose production is subject to wide variations in flow and composition.

In offshore oil and gas design only a small number of test/appraisal wells are drilled, and good engineering judgment has to be applied to evaluate the impact of the uncertainty of the data obtained from well fluid analysis on production, operation, and flow assurance.

Even bought-in chemicals have a specified range of properties, rather than a single value. There is in short no such thing as steady state.

We need to construct a range of realistic design scenarios, and our designs need to work in all of them. It may be that our provision for less probable scenarios has a shorter service life or requires more operator attention than that for the normal running case, but the plant has to be safe and operable under all foreseeable conditions.

Summary statistics

In real design scenarios you often have too little data to generate statistically significant design limits. We get a feel for the data by generating summary statistics; means, maxima, minima, confidence intervals, and so on.

Excel used to come with a plugin which helpfully generated this set of stats but now you need to plug in the formulae and functions yourself, although it doesn't take very long.

The lack of a formally statistically valid data set doesn't usually mean you get out of designing the plant. Sensitivity analysis can be used to see how much it matters if your data is unrepresentative.

The all-too-likely lack of rigorously valid data as a design basis should be something the designer is conscious of throughout the design. Those who claim that arrogance in process design can usefully be measured in nanomorans may be surprised to see me use the word, but we need to demonstrate humility.

For example, the well samples discussed in the last section are limited in number, and will not be fully representative for many reasons. Types of sampling methods, reservoir condition, insufficient well conditioning prior to sampling collection, inappropriate sampling collection methods, limitation of laboratory testing, contamination of sample by drilling mud, methanol, and so on may contribute to error. Different laboratories have been shown to produce wide inconsistency in compositional analysis even for the same sample.

All these uncertainties may affect the surface facility design further downstream such as pipelines and onshore terminals. Among the examples of impacts are incorrect sizing of separator, compressor, pumps; over-prediction or under-prediction of liquid holdup in pipelines, slug catcher capacity, fuel gas systems, and so on.

It is therefore important that the design parameters are not too tight and that a sufficient design margin is provided.

IMPLICATIONS OF NEW DESIGN TOOLS

Computer-based tools allow you to do more brute force calculations, so you don't have to design a plant in such a way as to make analysis simple. You should not, however, take advantage of this capability—your professional responsibility is to understand your plant. You also need to be careful not to be carried away by the precise looking outputs of these programs.

You should not use modeling and simulation programs as a substitute for design—by the time you have taken your usually rather flimsy design data and run it through a black box program written by someone who has never designed a plant, outputs are at best merely informative, and can easily be highly misleading.

There is more specific comment on these issues elsewhere in the book, but many of the things you were taught to do in university are not really design, and many of the tools you used are not used by professionals, with good reason.

IMPORTANCE OF UNDERSTANDING YOUR DESIGN

Recent stock market crashes have been at least contributed to by automated stock market modeling software, in many cases written by physical scientists and engineers based on models derived from the physical sciences very similar to process simulation software.

Your calculations, whether they be done by hand, in a spreadsheet, or a simulation program, are a model of the proposed system. It is infinitely better to have a simpler model which you understand well than a more complex one which you don't really understand.

In Henry Petroski's books, many of his engineering disasters come about as a result of people who thought they understood well-established design methodologies cutting safety margins and applying techniques to areas where their underpinning assumptions did not hold.

My advice is not to take responsibility for anything you don't understand well enough. This can make you unpopular if there is a lot of time pressure to get calculations signed off, but engineering is hard, and the stakes are high.

There's no shame in asking someone to show you why they think they have a design nailed down if you can't see how they have, and our professional responsibility means we have to imagine defending our actions in court later. Don't allow yourself to be pushed around by management. Their aims may be different from yours, which leads to a necessary tension in design.

MANAGER/ENGINEER TENSIONS IN DESIGN

Managers and engineers have to some extent different pressures upon them and different aims to meet during the design process, and the inevitable tension arising from this needs to be managed. An old engineering joke illustrates the differences in outlook:

A man in a hot air balloon realized he was lost. He reduced altitude and spotted a man below. He descended a bit more and shouted, "Excuse me, can you help me? I promised a friend I would meet him half an hour ago, but I don't know where I am."

The man below replied, "You are in a hot air balloon hovering approximately 30 feet about the ground. You are at approximately 53 degrees north latitude, and at 1 degree 13 minutes west longitude from the Greenwich meridian."

"You must be an engineer," said the balloonist.

"I am," replied the man, "but how did you know?"

"Well," answered the balloonist, "everything you told me is technically correct, but I have no idea what to make of your information, and the fact is I am still lost."

The man below responded, "You must be a manager."

"I am," replied the balloonist, "how did you know?"

"Well," said the man, "you don't know where you are or where you are going. You made a promise which you have no idea how to keep, and you expect me to solve your problem. The fact is you are exactly in the same position you were in before we met, but now, somehow, it's my fault."

Manager/engineer tensions I: risk aversion

There is a tension between external design consultants and product managers in client organizations, but even when designer and product manager are within the same organization, there may be a tension. Designers are usually risk averse, and management are often more risk tolerant.

In essence, management usually want to get a product to market as soon as possible, and with the absolute minimum possible margins of safety, and highest possible profit margin. Designers don't want to design things which don't work, they have a good idea of the limits of their analytical techniques, and they know that all design relies on approximation and heuristics.

When nondesigners look at our calculations and drawings, it all looks very mathematical, very sharp-edged, but these are precise-looking calculations of approximate values, and those straight lines on the drawing might be different in ten thousand ways.

I hear that management are now asking engineers to justify adding any margins of safety at all to designs done substantially using modeling programs, but that engineers are then being asked to carry out debottlenecking of plants which have been designed in this way.

So there will always be a tension between engineers and management. The Scotty Principle addresses this tension with respect to timescales:

Scotty Principle (n.) The de facto gold star standard for delivering products and/or services within a projected timeframe. Derived from the original Star Trek series wherein Lt. Cmdr. Montgomery "Scotty" Scott consistently made the seemingly impossible happen just in time to save the crew of the Enterprise from disaster.

1. *Calculate average required time for completion of given task.*

2. *Depending on importance of task, add 25–50% additional time to original estimate.*

3. *Report and commit to inflated time estimate with superiors, clients, etc.*

4. *Under optimal conditions the task is completed closer to the original time estimate vs. the inflated delivery time expected by those waiting.*

Urban Dictionary

However, margins of safety are not just about gaining a Scotty-style reputation as a miracle worker. When things are uncertain, designers should err on the side of caution, "under-promise and over-deliver." Occasionally you will lose out to those who are willing to gamble, but winning in that way makes them lucky, not good. Let's not trust to luck when so much is at stake.

Manager/engineer tensions II: the Iron Triangle

One other point of potential conflict between designers and management is covered by a trilemma sometimes known as the Iron Triangle; "*Fast, Good, or Cheap—pick any two.*"

It is common in my experience for managers and clients to want three out of three, but it just can't be done. Scotty knew about this too: "*I cannae change the laws of physics, Jim*"—neither can we change the iron laws of design.

WHOLE-SYSTEM DESIGN METHODOLOGY

Our initial conceptual design process needs to do more than just select broad technologies—such a decision making exercise on its own isn't really engineering design at all. A conceptual design needs to give us an approximate size and shape of unit operations and associated pipes, pumps, and so on.

They need to allow us to price at least all Main Plant Items (a term used in approximate costing functionally close to identical with Unit Operations). We need to lay all the kit out on a General Arrangement (GA) drawing to make sure it will fit on the site available. Ideally we would have some idea of how long it will take to build, and what it will cost to run it.

The sizing of unit operations for this purpose will be based on easy rules of thumb. We only need to know roughly how big they are. We will need to do a rough mass and possibly energy balance to allow us to size the units, as recycles can make flows far larger than are obvious by simple inspection.

We cannot optimize this rough design, and we should not try to. Conceptual design should under-promise and over-deliver. We should make very conservative assumptions, and err consistently on the side of caution. Things always take longer than you think, cost more, and have problems which are not immediately apparent. Save a bit of fat for later—you'll need it.

Detailed designs will need to address a number of issues which were disclosed but not resolved by the conceptual design stage. Rather than looking to scientific literature or modeling programs for answers, professional process plant designers usually look to the know-how of experienced engineers and equipment suppliers.

They may need to carry out slightly more rigorous design than at the conceptual stage to do this, and they will certainly need to get their P&ID, mass and energy balances tightened up considerably to do so. Again, this tightening-up should be based upon commercially available kit.

At this stage (if not earlier), the designer will need to supply the other engineering disciplines involved with the information they require to carry out their design.

Civil engineering designers will need to be provided with equipment weights, locations, and so on. Electrical engineers will need motor sizes and preferred starter types. Software engineers will need a control philosophy and a P&ID. They will all usually be in a hurry to get this information, but they will complain if it changes too many times during the design process.

The other disciplines (and sometimes the client) will come back with suggestions for how the overall design might be changed in ways which make it better, cheaper, and easier to build (or whatever). The plant designer needs to evaluate, negotiate and, fairly frequently, modify their process design to accommodate these changes.

Design for construction has to specify every detail, right down to the numbers and types of bolts used to connect pipe flanges. This stage is not merely the dull scheduling and documentation stage envisaged by theorists. Many design challenges remain—the devil is in the detail.

DESIGN STAGES IN A NUTSHELL

These are generalized consensus practice; additional disciplines and deliverables feature in certain sectors—see "Variations on a theme" to follow.

Conceptual design

- Construct Design Envelope.
- Produce Process Flow Diagram (PFD), P&ID, GA, unit op design, mass and energy balance, rough costing, and preliminary safety study.

  Purpose: To identify the design philosophy, the most promising technologies, rough cost, and footprint, then select the one or two most promising technologies.

  See the conceptual design section of later chapters for details of safety, costing, layout, materials and equipment selection, process control, and hydraulics aspects.

Intermediate design

- Produce more detailed PFD, P&ID, GA, unit operation design, mass and energy balance, control philosophy, detailed costing, robust safety study, obtain quotations and specifications from suppliers for equipment, and have mechanical, civil, control, and electrical design progressed in tandem.

Purpose: Choose sub-technologies, design approaches produce good quality costing and layout to allow evaluation and quite possibly to place an order for construction for long lead time items such as compressors.

See the intermediate design section of later chapters for details of safety, costing; layout, materials and equipment selection, process control, hydraulics, and optimization aspects.

Design for construction

- To produce the PFD, P&ID, GA, unit operation design, mass and energy balance, control philosophy, detailed costing, robust safety study, datasheets, schedules, obtain exact specifications and prices for equipment to be purchased, finalize technical bid evaluation, and have mechanical, civil, control, and electrical design progressed in tandem.

  Purpose: Design of plant to be built.

VARIATIONS ON A THEME

I was going to include here a big diagram showing all the stages of an ideal design methodology, but that might give you the impression there was such a thing. The truth is that the contents of the preceding chapters represent a core common approach, but that there are great variations between sectors, between companies, and between countries in a number of key issues, even if we stick to the whole-plant "grassroots" design which this book is about.

Operating companies, contracting/Engineering, Procurement and Construction (EPC) companies, and consultants of various kinds will all do a thing which they call design. The meaning of the term will vary, as will their understanding of terms like conceptual design, detailed design, and so on.

The job titles of the people involved in design, the tools they use, and the deliverables they produce will differ. There are many additional sector-specific deliverables I have not covered here, in the interests of simplicity. There are also sector-specific roles such as the petrochemical industry's control/instrument engineer.

The contents of the preceding chapters are true to the best of my knowledge for all sectors, as far as they go. There is, however, an infinite amount of detail which I have "omitted for clarity" as we engineers say.

FURTHER READING

Anon, 2012. Specification for Unfired, Fusion Welded Pressure Vessels. PD5500 British Standards Institute.

Anon, 2002. BS EN 13445: European Standard for Unfired Pressure Vessels. British Standards Institute, London.

Koolen, J.L.A., 2001. Design of Simple and Robust Process Plants. Wiley-VCH, Weinheim.

CHAPTER 8

How to Do a Mass and Energy Balance

INTRODUCTION

Most universities teach students to carry out mass and energy balances in single steady state scenarios in order to simplify the process for beginners. In the real world there is, however, no such thing as steady state.

Real plants are dynamic, with variable flows, compositions, temperatures, pressures, and so on. They have to produce product(s) to specification under all of these conditions, though it should be noted that "to specification" also implies a range of acceptable compositions rather than a single one.

Producing a model which accurately and dynamically models the operation of a plant which has not yet been constructed is impossible. We can, however, produce a number of steady state models across the design envelope, which allow us to generate performance curves and sensitivity analyses which give us confidence that the design will work (where the meaning of "work" is not perfection, but that implied by the brief) across all reasonably foreseeable conditions.

We need to exercise engineering judgment in deciding which scenarios we need to consider. In designing a water treatment plant we might, for example, look as a minimum at high flow and low flow scenarios, permutated with high contaminant and low contaminant levels.

We would also consider scenarios in which key process equipment is out of service for backwashing or maintenance, variations in feed temperature, degradation in dosed chemical quality, and so on.

This is almost always done in practice using a spreadsheet program, and the required permutation of scenarios becomes quite easy to achieve (and just as importantly, verify) if we structure our spreadsheet in a certain way.

Unsteady state

If you have an opportunity to view the "trends" screens on a supervisory control and data acquisition (SCADA) system, the dynamic nature of plant operation becomes very obvious (Figure 8.1).

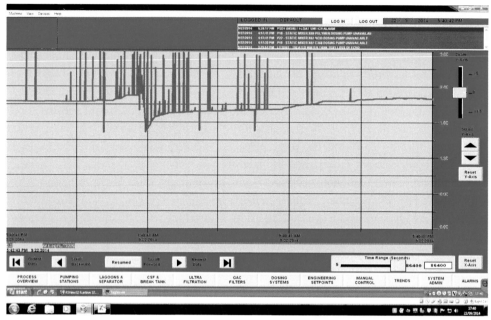

Figure 8.1 Unsteady state SCADA screen. *Image reproduced courtesy of the Process Engineering Group, SLR Consulting Ltd.*

Everything varies with respect to time on a process plant at some scale of resolution. Feedstock quality varies, as does product quality. Plant throughput varies both because we call for a different flow rate from the system, and because equipment degrades over service intervals, so that the same control inputs may produce different responses over time.

Other parameters vary for similar or different reasons. When we take off our steady state "training wheels," we need to understand the range of possible values which we can encounter in all parameters. A robust design works under all of these conditions.

It is often the case that the range of parameters we consider has an associated probability. If we have been provided with data which we are to use as a basis for design, the upper and lower limits of confidence intervals may well carry more weight than the maximum and minimum figures in the data.

So we are (consciously or not) designing a plant that will only probably work. Usually the required probability of success is implied by the performance tests we have to pass to have the plant accepted (though our design's probability of success should always be higher than that required by the test).

Implications of feedstock and product specifications

Most of things I have designed which have actually been built have been water and effluent treatment plants. The feedstock for such plants is a very variable flow of water with very variable levels of contamination. Drinking water plants ramp up and down production to match demand. Sewage and industrial effluent treatment plants have to reliably treat the often highly variable flows and compositions they are given.

Drinking water has to meet very tight and absolute maximum allowable value specifications for more than a hundred parameters. Effluent treatment product specifications are less complex, and may be probabilistic—95% compliance with specification might be fine.

What is true in water is true in all sectors: a plant needs to handle the worst feedstock it might encounter as well as the specification says it has to, producing the product to the specification required as reliably as is required.

There is no benefit in exceeding the plant's required performance, and it costs money to do so, but it has to meet required performance as specified, or for practical reasons a little better.

Stages of plant life

When setting out our design scenarios, we need to consider all stages of the plant's life. This is especially important if we have attempted to design in integration of systems.

If there is a stage of the plant's life, such as commissioning, start-up, shutdown, or maintenance, when we will need to run the plant in a different way, or need additional services, the full implications of this need to be considered in our mass and energy balance.

In the petrochemical industry, they mount campaigns called turnarounds where staff carry out more or less all the maintenance for a plant in a short sharp program. Is this how your plant will be maintained, or will it be little and often? The designer needs to know, and to take this into consideration from the stage of initial mass balance generation.

HANDLING RECYCLES

It is frequently the case that there are process streams on a plant which are returned directly or indirectly to the feed of the unit operation which they came from, and these are known generically as recycles.

As their introduction modifies the stream going into the unit operation, and the product of an operation is almost always affected by its feed stream, recycles affect themselves and, having been affected, affect themselves further, and so on. This may not be obvious during "steady state" operation, but it will be obvious to the designer.

Back when I was a student, we used to have to resolve this issue using iterative calculations done by hand. This was very dull indeed, the time-consuming nature of the approach meaning that nested recycles (recycles within recycles) were avoided.

Now we have MS Excel, which offers us a number of ways to carry out iterative calculations in seconds, and without the mistakes borne out of loss of attention which used to appear in hand calculations.

HOW TO SET IT OUT IN EXCEL

Here is how I set it out, which is pretty similar to the way most other professionals do. You don't have to do it the way I do, but you do need to address the problems I have done at least as robustly as my method does. Much of what professional engineers do habitually is intended to systematically avoid error, and make it easy to spot any residual errors.

Mass and energy balances like these are pretty complicated process models, which are very hard to hold in your head as a completely unified whole. Even the best examples will need rigorous double-checking by a second competent engineer, and you should assume that there will be mistakes before such checking.

I like to make my Excel spreadsheet look like the engineer's calculation pads we used to use back when we did hand calculations, and I teach my students to do this as well. If I do not do this, I get given spreadsheets which are hard to follow, with bits of calculation over in some obscure corner of the spreadsheet which no one notices.

Forcing the calculations into a succession of virtual sequential A4 sheets makes it easy to set out an annotated logical argument, and to follow the argument being made. So I recommend a vertical stack of these virtual pages be set out on each Excel tab.

I start with a header tab which sets out the given and assumed design parameters and gives an overview of the whole spreadsheet. All uses of these design parameters throughout the spreadsheet should be linked directly back to this header page. This makes it easy to vary the parameters to generate different scenarios.

If the cells containing the parameters have been labeled with descriptive names, the designer and the checker will not need to keep flicking back to the header to see what a parameter is when it is encountered on other sheets. I would in fact recommend that all cells whose values are copied across into calculations are so labeled to reduce errors and facilitate checking.

Each subsequent tab has a stack of pages which represent the design of a unit operation. I like to follow the order of the process flow on the Piping and Instrumentation Diagram (P&ID) in my virtual stack of Excel tabs. The mass and energy flows out of each unit operation should be (with the exception of those requiring the breaking of a recycle to avoid circular calculation errors—see later) directly linked to the inputs to the next operation.

This makes the spreadsheet capable of self-modifying to handle the various scenarios which it will be used to balance, removing the possibility that the designer will forget to change cells.

Each of these tabs will have what I call "checksums", calculations arranged so that they will be zero if the unit operation's mass balance is correct. These checksums will be carried across to the header tab, where they will appear as a single table, along with checksums for mass and energy balance checks at scales above the single unit operation.

Behind these mass and energy balance/unit operation sizing tabs I would usually have a few sheets of hydraulic calculations, so that I can include dynamic pump calculations based on the flows from the mass and energy balance in the spreadsheet for convenience. I have standard single-tab Excel spreadsheets (verified by an independent third party and then locked) which I insert for this purpose.

I also have such standard spreadsheet tabs for the more common unit operations which I design, saving time and reducing errors. Going to the trouble of producing such standard spreadsheets and having them validated is, however, only worthwhile if you are going to use them many times.

USING EXCEL FOR ITERATIVE CALCULATIONS: "GOAL SEEK" AND "SOLVER"

Microsoft Excel has a suite of commands known as "what-if analysis" tools. Process plant designers find two of these particularly helpful: Goal Seek and Solver.

Goal Seek allows us to vary the number in a spreadsheet cell until the value of a cell whose contents are calculated from the first cell's value is a number we specify.

Solver is more sophisticated, and allows us to minimize or maximize a value in a target cell.

These are very handy to plant designers—I use Goal Seek for the following purposes:

- Resolving recycles in mass and energy balances—use Goal Seek to make the difference between two mass balance formulae equal zero, and you have resolved the recycle;
- Dealing with iterative calculations, especially those with too many unknowns—I use this a lot for hydraulic calculations based on the Darcy—Weisbach equation and Colebrook—White approximation, as described in Chapter 9.

Solver, on the other hand, is well suited for optimization exercises such as sensitivity analysis and reactor design problems.

It may be useful to break the chain of calculations in nested recycle calculations with a cell whose value can be set manually, prior to the "what-if" command being run. This avoids "circular argument" error messages, but great care should be taken that all results yielded by such an exercise are sensible, as there is a greater probability of nonsensical results from this technique.

CHAPTER 9

How to Do Hydraulic Calculations

INTRODUCTION

I am an old man now, and when I die and go to Heaven there are two matters on which I hope for enlightenment. One is quantum electrodynamics and the other is the turbulent motion of fluids. And about the former I am rather more optimistic.

Sir Horace Lamb

At university, all chemical engineers study fluid mechanics, which is a kind of applied mathematics, usually combined with a bit of applied dishonesty.

The truth is that no one really understands the turbulent motion of fluids, or can predict it with a high degree of precision. Consequently, even the most basic fluid mechanics courses have to handle a transition from first principles mathematics to the rough heuristic of the Moody diagram.

This is the point where an honest lecturer should admit we can't actually use Bernoulli's equation to solve any useful problems, and we are consequently bringing in a chart based on empirical relationships determined by experiment to fill the gaps in our understanding with a fiddle factor.

Not all lecturers are so honest or insightful and students may leave university thinking that they understand something which no one does—until someone asks them to size a pump.

Hydraulics is the more practical cousin of fluid mechanics, which we mainly use to specify pump and equipment sizes as accurately as required for practical purposes. Engineers don't have to completely understand things in order to exercise sufficient control to achieve a given aim.

MATCHING DESIGN RIGOR WITH STAGE OF DESIGN

Right at the start of a design project, we need to know, for example, whether we intend to move fluids around our plant by gravity, or by pumping. Even to do the most basic layout we need to know approximate pipe diameters.

The nature of hydraulic calculations is essentially iterative. We need to have a pipe diameter and pipe length to work out the headloss down a pipe. We need to know the headloss to know the economic pipe diameter. We need to know static and dynamic head to know how big our pump needs to be. We need to know how physically large our pump is to know how long the pipe needs to be.

An Applied Guide to Process and Plant Design

115

We have to start this circular process somewhere. The method used by most process engineers is described in the following sections. Levels 1/2 are used at conceptual design stage, Level 2/3 at detailed design. Level 4 is used only in rare cases of complexity.

Hydraulic calculations have three main components: static head (the elevation from reservoir to point of discharge, plus any atmospheric pressure difference between reservoir and point of discharge), straight run headloss (headloss due to friction at operating flowrate due to straight pipe sections), and fittings headloss (headloss due to friction at operating flowrate due to bends, tees, valves, etc.).

In water process engineering, we make a fair amount of use of open channels, and I consequently have to do quite a lot of open channel hydraulics. However, I am going to leave that out of this book, as it can be quite complicated and there isn't very much of it in most other process sectors.

I am also going to leave out calculations for multiphase flow and water hammer. I suggest you look at Donald Woods' book (see Further Reading) for guidance on shortcut methods for these though, in practice, you might want to call in a specialist if your rough calculations make you think these conditions likely.

Level 1—superficial velocity

Superficial velocity is the same thing as average velocity (i.e., the volumetric flowrate in m^3/s divided by the pipe's internal cross-sectional area in m^2—its units are in this example m/s).

A very quick way of starting our hydraulic calculations is to use the following rules of thumb from acceptable superficial velocities:

- Pumped water-like fluids <1.5 m/s
- Gravity fed water-like fluids <1 m/s
- Water-like fluids with settleable solids >1, <1.5 m/s
- Air-like gases 20 m/s

Two-phase flow is hard to predict, and should be designed out if at all possible—headlosses can be one thousand times that for single phase flow.

These rules will usually give sensible headlosses for the sort of pipe lengths normally found on process plants.

Level 2—nomograms, etc.

The most difficult part of a headloss calculation is determining the straight run headloss. It isn't really that difficult, but we have to do it many times, so a quick method is handy at earlier stages of the design.

Liquids

Pipe manufacturers and others produce tables and nomograms which can be used to quickly look up headloss due to friction for liquids (Figure 9.1).

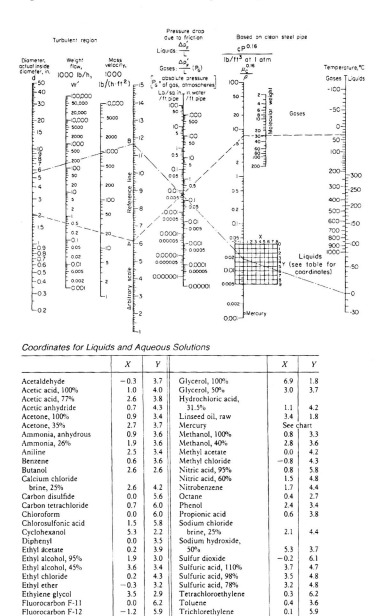

Coordinates for Liquids and Aqueous Solutions

	X	Y		X	Y
Acetaldehyde	−0.3	3.7	Glycerol, 100%	6.9	1.8
Acetic acid, 100%	1.0	4.0	Glycerol, 50%	3.0	3.7
Acetic acid, 77%	2.6	3.8	Hydrochloric acid,		
Acetic anhydride	0.7	4.3	31.5%	1.1	4.2
Acetone, 100%	0.9	3.4	Linseed oil, raw	3.4	1.8
Acetone, 35%	2.7	3.7	Mercury	See chart	
Ammonia, anhydrous	0.9	3.6	Methanol, 100%	0.8	3.3
Ammonia, 26%	1.9	3.6	Methanol, 40%	2.8	3.6
Aniline	2.5	3.4	Methyl acetate	0.0	4.2
Benzene	0.6	3.6	Methyl chloride	−0.8	4.3
Butanol	2.6	2.6	Nitric acid, 95%	0.8	5.8
Calcium chloride			Nitric acid, 60%	1.5	4.8
brine, 25%	2.6	4.2	Nitrobenzene	1.7	4.4
Carbon disulfide	0.0	5.6	Octane	0.4	2.7
Carbon tetrachloride	0.7	6.0	Phenol	2.4	3.4
Chloroform	0.0	6.0	Propionic acid	0.6	3.8
Chlorosulfonic acid	1.5	5.8	Sodium chloride		
Cyclohexanol	5.3	2.2	brine, 25%	2.1	4.4
Diphenyl	0.0	3.5	Sodium hydroxide,		
Ethyl acetate	0.2	3.9	50%	5.3	3.7
Ethyl alcohol, 95%	1.9	3.0	Sulfur dioxide	−0.2	6.1
Ethyl alcohol, 45%	3.6	3.4	Sulfuric acid, 110%	3.7	4.7
Ethyl chloride	0.2	4.3	Sulfuric acid, 98%	3.5	4.8
Ethyl ether	−0.3	3.2	Sulfuric acid, 78%	3.2	4.8
Ethylene glycol	3.5	2.9	Tetrachloroethylene	0.3	6.2
Fluorocarbon F-11	0.0	6.2	Toluene	0.4	3.6
Fluorocarbon F-12	−1.2	5.9	Trichlorethylene	0.1	5.9
Fluorocarbon F-21	−0.4	5.9	Turpentine	1.1	3.1
Fluorocarbon F-22	−1.7	5.5	Vinyl acetate	0.4	4.2
Fluorocarbon F-113	0.9	6.2	Water	2.0	4.2
Formic acid	1.5	4.5			

Figure 9.1 Pipe flow chart nomogram. *Copyright material reproduced from Sandler, H.J. and Luckiewicz, E.T. (1987) Practical Process Engineering: A Working Approach to Plant Design.*

We can then calculate fittings headloss by the *k* value or equivalent diameter method (obtaining a count of valves etc. from the Piping and Instrumentation Diagram (P&ID), and bends, tees, and so on from the General Arrangement (GA) drawing), and work out the static head from heights measurable from our GA, plus vessel pressures read from our Process Flow Diagram (PFD). This is one of the reasons why even quite early stage designs need to produce all three of these drawings.

It may be seen that once we have carried out the hydraulic calculations, our pump and possibly pipe sizes will need to change, as might minimum and maximum operating pressures at certain points in the system. There might even be a requirement to change from one pump type to another, or to change from a fan to a blower or from a blower to a compressor.

So there is a stage of design development which takes a set of preliminary drawings and modifies them to match likely hydraulic conditions across the design envelope. This stage requires us to do lots of approximate hydraulic calculations before the design has settled into a plausible form.

We consequently do the quickest and the least rigorous calculations which meet the needs of this stage of design development as described in this section.

Net positive suction head

At Level 2, I would also recommend calculating net positive suction head (NPSH), as it can affect a lot more than just pump specification. There is a good description of what this is and how to do it in Coulson and Richardson Volume 1. I recommend producing an Excel spreadsheet based on this approach, using the Antoine equation to estimate vapor pressures.

Note that NPSH is calculated differently for centrifugal and positive displacement pumps, varying with pump speed for positive displacement (PD) rather than pressure as with centrifugal.

Gases

If we are working with an air-like gas, we can use the charts of friction losses in ducts for air which are readily available to estimate straight run headloss (Figure 9.2).

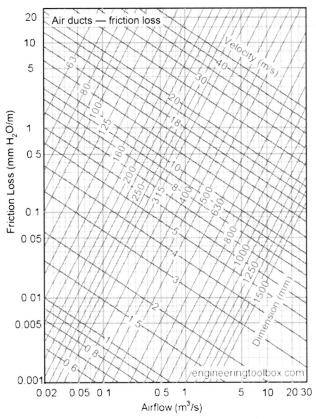

Figure 9.2 Gases: duct chart. *Copyright material reproduced courtesy of www.engineeringtoolbox.com.*

If headloss due to friction is <40% of upstream pressure (as it usually is), we can ignore compressibility effects for gases at Level 2, and use the same method as suggested for liquids above.

Level 3 (now superseded)—Moody diagram

Students are mostly taught to calculate straight run headloss using a Moody diagram, which is a summary of empirical experiments (and essentially an admission of defeat on the part of the mathematicians and scientists responsible for fluid mechanics—they couldn't make their sums work without these fiddle factors taken from experimental data).

The Moody diagram is one of the things superseded by MS Excel. As Excel can't read charts, we use curve-fitting equations which approximate the Moody diagram's output.

While this is an approximation, it might well be closer to the true experimental value than is read by the average person from an A4 copy of a Moody chart. In any case, it's a fiddle factor.

Level 3 (updated): spreadsheet method

Liquids

I personally use the Colebrook—White approximation to give me the fiddle factor which I would once have read from the Moody diagram, and I plug this into the Darcy—Wiesbach equation to work out straight run headloss, with an iterative method based on Excel's Goal Seek function which I cover in Chapter 8.

I recently read a paper (see Further Reading) which suggested there are new and more accurate curve fitting equations, and I might have got around to modifying my standard hydraulic calculation spreadsheet if I hadn't gone to all the trouble of having it validated.

So if you are producing your own spreadsheet for this purpose, I suggest you look into the Zigrang and Sylvester or Haaland's equations, which this paper recommends, to generate your fiddle factor.

This approach allows you to calculate straight run headloss to the degree of accuracy required for more or less any practical application.

Static head and fittings headloss can then be calculated as in Level 2, and it can all be added up to generate a delivery side headloss.

Suction side headloss and NPSH should also be calculated, and all of this used to generate an approximate pump power rating for a centrifugal pump using the equation:

$$P = Q\rho gh/\left(3.6 \times 10^6\right)\eta$$

where

P = power (kW)
Q = flowrate (m³/h)
ρ = density of fluid (kg/m³)
g = acceleration due to gravity (9.81 m/s²)
h = total pump head (m of fluid)
η = pump efficiency (allow 0.7 if you don't have a figure).

The manufacturer will give you the precise power ratings and motor size, but the electrical engineers will need an approximate value of this (and pump location) quite early on in the design process, to allow them to size their power cables. You should err on the side of caution in this rating calculation, as the electrical engineers will be a lot happier with you if you come back later to ask for a lower power rating than if you ask for a higher one.

Gases

Compressibility can make this all get a bit complex, but we can simplify matters. Crane, the valve manufacturers, proposed a simplified method in a technical paper first published in 1942 (see Further Reading).

If headloss is <40% of upstream pressure (as it usually is), gas compressibility can be ignored, and we can use the Darcy—Weisbach equation/k-value method to determine headloss. The gas density used should be consistently either that upstream or downstream for headloss <10% of upstream pressure.

For headloss of 10−40% of upstream pressure, use the density at the average of upstream and downstream conditions. If it is >40% of upstream headloss, we will need to consider compressibility and use the Weymouth, Panhandle A, and Panhandle B equations. It should be clear that this will require an iterative design process.

Level 4—CFD

I have never had to carry out a Computational Fluid Dynamics (CFD) study, though I know other professional engineers who have, so it isn't ridiculously theoretical and impractical.

It is, however, rare enough that it is more likely that you will go out to a specialist subcontractor to do it for you rather than buy the software and learn to use and validate it for a one-off exercise.

HYDRAULIC NETWORKS

The previous section is about how to calculate the headloss through a single line, but what about the common situation where we have branched lines, manifolds, and so on?

Each branch is going to take a flow proportional to its headloss, and its headloss will be proportional to its flow. Producing an accurate model can become complex very quickly. Things which are at all hard to model/understand are generally not robust design unless this represents the only workable approach.

My approach to this is to reduce complexity and improve design as follows:
- Avoid arrangements of manifolds which give a straight-through path from feed line to branch. Entry perpendicular to branch direction is preferred.
- Oversize manifolds such that superficial velocity never exceeds 1 m/s at the highest anticipated flowrate.
- Step the manifold diameter along its length to accommodate lesser flows to further branches.
- Include a small hydraulic restriction such that branch headloss is 10−100 times that from one end of the manifold to another.
- Design in passive flow equalization throughout the piping system wherever possible by making branches hydraulically equivalent.

I then do headloss calculations for each section of the plant at expected flows to find the highest headloss flow path through my simplified design.

I use this path to work out the required pump duty. I will test it at both average flow with working flow equalization, and at full flow through a single branch. Usually these don't differ all that much, and as I know that the more rigorous answer lies between them, I don't worry about it. Only if the results of this approach seem problematic will I do a more rigorous analysis.

To do a more rigorous analysis, I create an Excel spreadsheet based on the Hardy Cross method to solve for individual pipe flows. The "Solver" function can be used to find the change in flow which gives zero loop headloss.

There are many computer programs available to do these calculations, but I would personally always rather produce a simple model in MS Excel—which I completely understand—than use black box programs.

PUMP CURVES

A notable omission from university courses is an understanding of how to read a pump curve, which is an essential requirement to do what we are probably going to do with the head/flow pairs we calculated across the design envelope.

The most frequent use of pump curves is for the selection of centrifugal pumps, as the flow rate of these varies so dramatically with system pressure. Pump curves are used far less frequently for positive displacement pumps.

A basic pump curve plots the relationship between head and flow for a pump at a given supply frequency. On more sophisticated curves, there may be nested curves representing the flow/head relationship at different supply frequencies or rotational speeds, with different impellers, or different fluid densities. The pattern is that curves for larger impellers or faster rotation lie above smaller impellers or slower rotation, and lower specific gravity above high for centrifugal pumps.

Let's start with a basic curve (Figure 9.3):

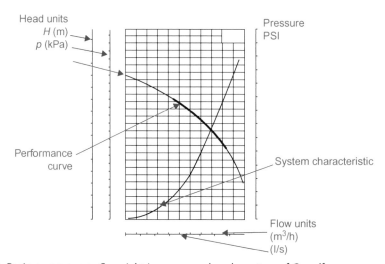

Figure 9.3 Basic pump curve. *Copyright image reproduced courtesy of Grundfos.*

Along the horizontal axis we have increasing flow (*Q*), and along the vertical axis, increasing pressure (*H*). The curve shows the measured relationship between these variables, so it is sometimes called a *Q/H curve*. The intersection of the curve with the vertical axis represents the *closed valve head* of the pump. These pumps are generated under shop conditions and ideally represent average values for a representative sample of pumps.

We can use our calculated flow/head pairs to plot a system head on the same axes, and see where our system head meets the *Q/H* curve. This will represent the operating or *duty point* of the pump.

We will have a system head curve for the expected range of flows at a given system configuration. Throttling the system will give a different system curve. We will need to produce a set of curves which represent expected operating conditions, generating a set of duty points.

That's it as far as our basic curve is concerned, but it is common to have efficiency and motor rating curves plotted on the same graph (but not the same vertical axes) as in the example in Figure 9.4.

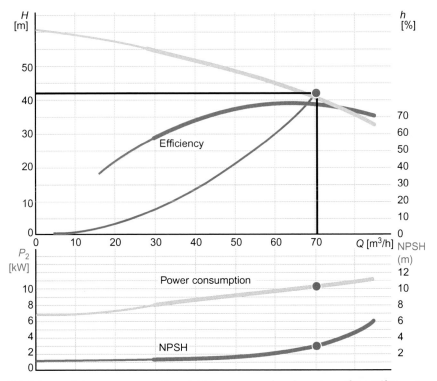

Figure 9.4 Intermediate pump curve. *Copyright image reproduced courtesy of Grundfos.*

So we can see that we can draw a line vertically from the duty point to the efficiency curve, and obtain the pump efficiency at the duty point by reading the vertical axis at the point of intersection. Similarly we can draw a vertical line to the motor duty curve, and obtain a motor power requirement.

Having tackled these basic and intermediate curves, we can look at the common format of professional curves, incorporating efficiency, NPSH, and impeller diameters like this (Figure 9.5).

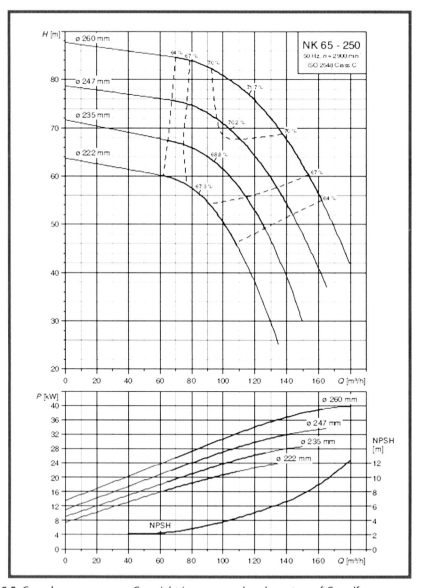

Figure 9.5 Complex pump curve. *Copyright image reproduced courtesy of Grundfos.*

These start to look a bit confusing, but the thing to bear in mind is that, just as with the simpler examples, the common axis is always the horizontal one of flowrate. So the corresponding value on any curve is vertically above or below the duty point.

These more advanced curves usually come with efficiency curves, and it is usually visually obvious that these curves seem to bound a region of highest efficiency. At the center of this region is the *best efficiency point* or BEP.

We will want to choose a pump which offers good efficiency across the range of expected operating conditions. Note that we are not necessarily concerned with the whole design envelope here—it is not crucial to have high efficiency across all conceivable conditions, just the normal range.

A well-selected pump will have a BEP close to the duty point. If the duty point is way over to the right of a pump curve, well away from the BEP, this is not the right pump for the job. Try another.

These are the basics of centrifugal pump selection. If you are in a position to influence which pump is purchased, any pump supplier's representative would be happy to talk to you about pump selection for as long as you are willing to listen. Probably buy you lunch too, though obviously that wouldn't affect your choices.

Even with the most cooperative pump supplier, the curves you want in order to make a pump selection may not be available, as is commonly the case when we want to use an inverter to control pump output by speed. We can, in this case, generate the required curves for ourselves using pump affinity relationships. The laws are:

- Flowrate$_2$/Flowrate$_1$ = Impeller diameter$_2$/Impeller diameter$_1$ = Pump Speed$_2$/Pump Speed$_1$
- Dynamic Head$_2$/Dynamic Head$_1$ = (Impeller diameter$_2$/Impeller diameter$_1$)^2 = (Pump Speed$_2$/Pump Speed$_1$)^2
- Power Rating$_2$/Power Rating$_1$ = (Impeller diameter$_2$/Impeller diameter$_1$)^3 = (Pump Speed$_2$/Pump Speed$_1$)^3
- NPSH$_2$/NPSH$_1$ = (Impeller diameter$_2$/Impeller diameter$_1$)^x = (Pump Speed$_2$/Pump Speed$_1$)^y

Where subscript 1 designates an initial condition on a known pump curve, and subscript 2 is some new condition.

The NPSH relationship is a lot more approximate than the others. x lies in the range -2.5 to $+1.5$, and y in $+1.5$ to $+2.5$. A worst-case estimate can be established using the maximum quoted x and y figures if impeller speed or diameter is to be increased, and the lowest figures if it is to be decreased.

FURTHER READING

Anon, 2009. Flow of Fluids through Valves, Fittings, and Pipe. Technical Paper 401, Crane Co.

Cross, H., 1936. Analysis of flow in networks of conduits or conductors. Engineering Experiment Station Bulletin No. 286.

Genić, S., Srbislav Genić, Arandjelović, I., Kolendić, P., Jarić, M., Budimir, N., Genić, V., 2011. A review of explicit approximations of Colebrook's equation. FME Transact. 39, 67−71.

Woods, D.R., 2007. Rules of Thumb in Engineering Practice. Wiley-VCH, Weinheim.

PART 3

Low Level Design

Enough background—exactly how do you design like a pro? Some of this section is the professional version of the subjects which are taught in university, and some will be completely new or may even contradict what is taught in academia.

CHAPTER 10

How to Design and Select Plant Components and Materials

INTRODUCTION

Engineering is the art of modeling materials we do not wholly understand, into shapes we cannot precisely analyze so as to withstand forces we cannot properly assess, in such a way that the public has no reason to suspect the extent of our ignorance.

A.R. Dykes

The selection of the basic subcomponents of process plants is an essential part of what plant designers do. There is often a fundamental misunderstanding in academia of what constitutes the elements of a process plant.

Process plants are not made of ideas, or (at an engineer's usual resolution of vision) even of chemicals. Process plants are made of commercially available products.

We are not usually employed to select process chemistry, but to specify the pumps and valves, pipes and tanks used to construct the plant in which those chemistries occur.

As a result of the composition of university departments, taught chemical engineering often overemphasizes science and mathematics to the point where graduates lack the broad overview of available technologies which allows them to make such a selection.

Such qualitative knowledge may seem less intellectually rigorous than science and mathematics, but it is actually far more sophisticated to exercise multidimensional judgment in a mental space of the qualities of various process options than to grind through a rote calculation which a computer could beat you at.

In this chapter I will attempt to offer the broad guidance on the selection of common items which is missing from many chemical engineering programs.

WHAT PROCESS ENGINEERS DESIGN

The essence of process engineering is integration of complex systems, but in order to integrate systems, the designer has to have some knowledge of the characteristics of those systems which affect integration.

To be more specific, certain types of materials, for example, are more suited to a given range of pressures, temperatures, and chemical and physical compositions than others. Matching the ranges of these parameters in the plant design envelope to suitable materials is usually thought of as the process plant designer's job. Similarly, the selection of pumps, heat exchangers, instrumentation, valves, and so on is usually thought to be part of process plant design.

The information required to make these selections is largely absent from chemical engineering degrees, justified by the idea that it is mere qualitative data which is insufficiently intellectually demanding for university level education.

Similarly, the qualitative criteria used to choose between separation processes and other technologies are frequently thought of as too shallow and easy to be worth a student's time. Students mostly concentrate on a few selected processes which can be used to illustrate scientific principles or mathematical techniques.

Practitioners understand that such knowledge is actually capable of forming the basis of quite subtle and sophisticated multidimensional reasoning, and that providing such judgments is one of the basic expectations of a process plant design engineer.

I will attempt in the following chapter to provide matrices showing a number of dimensions which may be used to choose between options for materials of construction, valves, pumps, blowers compressors and fans, separation processes, and heat exchangers. I will also offer information on specification of electrical components and instrumentation.

MATCHING DESIGN RIGOR WITH STAGE OF DESIGN

At conceptual design stage it is often important to know at a category level what kinds of components we are thinking of using. Whether we are going to use rotodynamic or positive displacement pumps, membranes or distillation, globe or butterfly valves, carbon steel or plastic is usually known to the process plant designer by the time their initial drawings are done. All of these decisions affect the fundamental characteristics of the design and have implications for cost, safety, and robustness.

Experienced professionals might not even know they are making some of these choices (which might be strongly affected by custom in the sector and personal preference), but the beginner has to make conscious choices. There may be little formal documentation of some of these choices at this early stage, but the designer has to make many of them to carry out even a conceptual design.

At the detailed design stage we are selecting specific commercially available items of equipment. We produce datasheets which set out the detailed specification of the item, or incorporate our choices in the case of materials of construction. Manufacturers may, on sight of these datasheets, feed back to us information which allows us to refine or reconsider our choices.

More generally, the more detailed analysis may show some of our conceptual design choices to be less than ideal, so we might change our minds. If someone else did the conceptual design, it can be a good idea to ask them why they chose the component which you want to change. They may have had a good reason to choose it based on factors you are unaware of. If that is not possible, the project design philosophies may be of assistance.

Design for construction generates a lot more detailed documentation, and experienced engineers will be likely to review your design choices before this is finalized. They may ask for changes based on their experience of what works, or more importantly still, what does not.

MATERIALS OF CONSTRUCTION

Plant designers need to know which materials are going to be suitable for the duty to which they intend to put the plant, as well as the duties to which it might (intentionally or unintentionally) be put.

This is not determined so much by material science as by practical experience, and a broad qualitative knowledge of available materials and their strengths and limitations.

There are also traditional default positions in various process sectors. For example, the generic pipe material is carbon steel in the oil and gas industry, and plastics in the water industry—even highly corrosive water is transported in carbon steel piping in the oil and gas industry.

Though this seems odd to a water specialist, this is not wrong, as long as suitable corrosion allowances are made, and the consequent increased metal ion content of the water is acceptable from a process point of view (Table 10.1).

Table 10.1 Materials of construction

	Relative price	Temperature rating	Pressure rating	Water resistance[a]	Organic solvent resistance	Acid resistance	Alkali resistance	Abrasion resistance	Chloride resistance	UV resistance	Hardness	Toughness
Metals												
Cast iron	L	H	M/H	M	H	L	L/M	H	L	VH	H	M/H
Carbon steels	L	H	H	L	H	L	L	M/H	L	VH	H	H
Stainless steels	M	H	VH	H	H	L	L/M[b]	VH	L/M[b]	VH	H	VH
Bronzes	M	M/H	H	L	H	L	L	M	M	VH	M	M
Brasses	M	H	H	H	H	L	VL	M	M	VH	M	M
Hastelloy	VH	VH	VH	VH	H	H	M	H	H	VH	H	H
Tantalum	VH	VH	M/H	VH	H	VH	M	H	H	VH	H	VH
Inconel	VH	VH	VVH	VH	H	M/H[b]	H	H	M/H[b]	VH	H	H
Aluminum	M/H	M/H	M/H	H	H	L	L	M	L	VH	H	H
Titanium	H	H	M/H	H	H	M	L	H	VH	VH	H	H
Precious metals	VVH	H	L/M	H	H	M/H	VH	L/M	VH	VH	L/M	L/M
Plastics												
PVC	L	VL/L[b]	M/H[b]	H	M[b]	H	H	M	H	VL	M	H
ABS	L	VL	L	H	M	H	H	M	H	M	M	H
PP/PE	L	VVL	VL	H	L	H	H	L/M	H	H	L	M
PS	L	VL	L/M	H	L	H	H	L	H	H	L	H
Acrylic	M	L	L/M	H		H	H	H	H	H		H
Nylon	M	L/M	M	M/H	L/M	M	H	H	H	H	M	H
Polybenzi-midazole	H	M/H	M	H	H	L/M	L	H	H	H	M	H
Fluoropolymers												
PTFE	H	M	VL	H	H	H	H	M	H	H	M/H	M
PVDF	VH	M	L/M	H	H	H	H	M	H	H	M/H	M/H
Other												
Ceramics	M	VVH	VVH	H	H	L/M	L	VH	H	VH	VH	L
Graphite	H	H	L/M	H	H	H	L/M	L	H	H	L	L
Glasses	M	M	L/M	H	VH	H	M	H	H	H	VH	L
Rubber	L	L	L/M	H	L/M	L	H	H	H	M	L	H
Composites	H	L	H	H	M/H	L/M	H	M	H	M	M	H

[a]Corrosive water by Langelier index (LSI).
[b]Varies by grade.

Scaling and corrosion

Prediction of the scaling or corrosive nature of fluids impacts on many areas of design from the selection of tube or shell side duty on a heat exchanger of a fluid to fouling factors and corrosion allowances. There is a lot to this subject, but I will confine myself to covering a few key aspects in outline.

Corrosion allowances are temperature, pressure, and internal and external chemical environment dependent, and limited life stipulations may be needed for pressure vessels in corrosive environments.

Effects such as erosion by entrained solids, galvanic corrosion, stress corrosion cracking, hydrogen embrittlement, cavitation, vibration, thermal expansion/contraction, and water hammer may all cause the premature failure of metal components.

New processes especially might need extensive testing of materials. Materials specialists should assess suitability of materials of construction at detailed design stage. For earlier stages of design, Table 10.2 (from "Practical Process Engineering") may be useful.

There are quite a number of indices used to predict whether water is going to be scaling or corrosive, which are all at best quite rough heuristics.

Table 10.2 Corrosion table

	Brass	Bronze	Alloy 20	Hastelloy C	Monel	304 stainless steel	316 stainless steel	Titanium	Silicon iron	Tantalum	Copper	Aluminum	Carbon steel	Butyl rubber	Epoxy	Hypalon	Natural rubber	Neoprene	Nitrile rubber	Nylon	Phenolic	Polyethylene	Polypropylene	PVC	Silicone elastomer	TFE	Ceramic, alumina	Graphite
Acetaldehyde	C	C	A	A	B		A	A	A	A			N		N	N	A	A		N			B	N	A	A	A	A
Acetic acid, 20%	C	C	A	A	B	A	A	A	A	A	A	A	N	A	A	A	C	C	A	N		C	A	A		A	A	A
Acetic acid, 80%	C	C	A	A	B		A	A	A	A	A	A	N	A	B	A	C	A	N	N	B	C	B	C		A	A	A
Acetic acid, glacial	C	C	B	A		A	B	A	A	A	A	A	C	C	B	N	C	C	N	N	A	A	B	N		A	A	A
Acetic anhydride	A	A	A	A	N	N	C	A		A	A	A	A	A		C		A		A	A	A	C	N	C	A	A	
Acetone	A	C	C	A	C	N	C	C	C	A	C	C	A	A	C	A	N	A	N	A	A	A	A	A	C	A	A	A
Aluminum chloride	C	C	C	A	C	C	C	A		A	C	C	N		A	A	N	A	N	N	A		A	A		A	A	A
Aluminum sulfate	N	C	C	A	C	C	C	A	C	A	C	C	N	A	A	A	N	A	N	A	A	A	A	A		A	A	
Ammonia, 10%	C	C	C	A	N	C	C	A		A	N	C	N		C	A	N	A	N	N			A	A		A	A	A
Ammonium chloride	N	N	A	A	C	C	C	A	C	A	N	C	C	A	A	A	A	A		A	A	A	A	A	A	A	A	A
Ammonium nitrate	N	N	A	A	N	A	A	A	C	A	N	C	C		A	A	N	A		N		A	A	A		A	A	A
Ammonium phosphate	N	N	A	A	N	A	C	A	C	A	N	C	C		C	N	N	A		N		A	A	A		A	A	A
Ammonium sulfate	N	N	A	A	C	A	A	A	C	A	C	A	C	A	B	C	B	A		C		A	A	A	A	A	A	A
Amyl acetate	B		A	A	A		A	A		A	A	A		A	B	C	N	A	N	A	A	A	A	A	A	A	A	A
Amyl alcohol	B		A	A	A	A	A	A	A	A	A	A			C	C	N	A	N	C	A	A	A	A		A	A	
Amyl chloride	N		A	A	N	N	B	A	C	A	N	N	A		N	N	B	A		A		N	N	N		A	A	A
Aniline	N		A	A	C	B	N	A	C	A	N	N	N		C	C	B	A		N		N	B	C		A	A	A
Aqua regia	N	N	N	A	N	B	A	A	C	A	N	N			B	N	B	N		N	B	B	N	N		A	A	A
Arsenic acid	N		A	A	A	B	B	A	N	A	N	N		A		C	N	N	N	A	A	B	N	N		A	A	A
Barium chloride	A	N	A	B	B	B	N	A	A	A	A	A	C	A	A	B	N	N	N	N		B	B	A	N	A	A	
Barium sulfate	A		A	A	A	B	B	A	A	A	A	A		A	B	C	B	A	N	N		B	B	A	N	A	A	A
Beer			A	A		A	A	A	A	A	A	N					N	A		A		A	A	A		A	A	A
Benzaldehyde	A		A	A	A	A	A	A	A	A	A	A	C	A	A	A	C	A	N	A	B	N	A	A	A	A	A	A
Benzene	A	A	A	A	A	C	A	A	A	A	A	A	A	A	A	A	N	N	N	N	A	C	A	A	N	A	A	A
Benzoic acid			A	A		A	A	A	A	A	A	A		A	A	N	N	N	N	C	A	C	A	A	N	A	A	A
Borax	A	C	A	A	C	C	A	B	A	A	A	N	A	A	A	N	N	N	N	N	A	A	A	A	N	A	A	A
Boric acid	N	A	A	A	A	A	A	A	A	A	C	A			A	A	A	A	N	C	A	C	A	A	N	A	A	A
Bromine water	C	A	N	A	C	N	A	A	A	A	C	N	N		B	N	A	N	N	N	A	A	A	B	N	C	A	N
Butyl acetate	A	C	C	A	A	C	C	A	N	A	A	B	N	A	B	N	N	N	N	A	N	N	N	N	N	A	A	A

(Continued)

	Brass	Bronze	Alloy 20	Hastelloy C	Monel	304 stainless steel	316 stainless steel	Titanium	Silicon iron	Tantalum	Copper	Aluminum	Carbon steel	Butyl rubber	Epoxy	Hypalon	Natural rubber	Neoprene	Nitrile rubber	Nylon	Phenolic	Polyethylene	Polypropylene	PVC	Silicone elastomer	TFE	Ceramic, alumina	Graphite
Butyric acid	B	A	A	A	A	C	A	B	A	A	A	B			C	Z	A	Z		Z		Z	A	C	Z	A	A	A
Calcium bisulfate	Z	A	C	A	Z	C	C	C	Z	A	A	A		A	A	A	Z	A	A		A	A	A	A	Z	A	A	A
Calcium chloride	C	A	C	A	C	C	A	A	A	A	C	C	C	A	A	A	C	A	A	Z	A	A	A	A		A	A	A
Calcium hypochlorite	A	C	C	C	A	C	C	A	A	A	A	A	C	A	B	A	B	Z	A	A	A	Z	A	A	C	A	A	A
Furfural	A	C	A	A	A	C	C	A	A	A	A	A	A	Z	B		B	Z	Z	A	A	Z	Z	Z	A	A	A	A
Gasoline	A	A	A	A	A	A	A	A	A	A	A	A	A	A	C	Z	Z	Z			A	Z	A	B		A	A	A
Glycerine	A	A	A	A	A	A	A	A	A	A	A	A	A		A	C	A	A		A	A	A	A	A	C	A	A	A
Heptane			A	A	Z	A	A	A	A	A		A		C	C	C	B	C	C		A	Z	B	A	Z	A	A	A
Hexane			A	A	C	A	A	A		A		A		C	B	C	B	A	C	A		A	B	B		A	B	A
Hydrobromic acid, 20%	Z	Z	Z	A	Z	Z	Z	A	Z	A	Z	Z	Z	Z	C	A	Z	A	Z	Z	A	A	A	A	C	A	B	A
Hydrochloric acid, 0–25%	Z	Z	Z	A	C	Z	Z	A	A	A	C	Z	Z	A	A	C	Z	C	Z	Z	A	A	A	A	Z	A	B	A
Hydrochloric acid, 25–37%	Z	Z	Z	A	C	Z	Z	A	A	A	C	Z	Z		C	C	Z	A	Z	Z		A	A	A		A	B	A
Hydrocyanic acid	Z	Z	A	A	C	A	A	A	A	A	Z	Z	Z	A	A	A	Z	A		Z	A	A	A	A	A	A	B	A
Hydrofluoric acid, 10%	Z	Z	B	A	A	A	Z	A	A	Z	Z	Z	Z	A	A	A	A	A	Z	Z		A	A	B	C	A	Z	A
Hydrofluoric acid, 30%	Z	Z	B	A	A	A	Z	A	A	Z	Z	Z	Z	A	A	C	A	A	Z	Z		A	A	B	Z	A	Z	A
Hydrofluoric acid, 60%	Z	Z	B	A	A	B	A	A	Z	Z	Z	Z	Z	A	A	A	A	A	Z	Z		A	A	C	A	A	Z	A
Hydrogen peroxide, 30%	Z	Z	A	A	A	B	B	A	Z	A	C	C	C	B	A	C	A	Z		Z	A	A	A	A	C	A	A	A
Hydrogen peroxide, 50%	Z	Z	A	A	A	B	B	A	A	A	C	C	C	Z	A	A	A	C	A	Z		Z	B	B	C	A	A	A
Hydrogen peroxide, 90%	Z	Z	A	A	A	C	A	Z	Z	A	C	C	C	A	A	C	Z	C	Z	Z	A	A	B	Z	A	A	Z	A
Hydrogen sulfide, aqueous	Z	Z	A	A	C	C	A	A	A	A	Z	C	C	Z		A	Z	A		Z		Z	B	B	C	A	Z	A
Iodine in alcohol	Z	Z	A	A	C	C	A	A	Z	A	C	C	C	A	A	C	Z	Z		Z		A	A	A	A	A	Z	A
Kerosene	A	Z	A	A	A	C	A	A	A	A	A	A		A		A	Z	A	Z	Z		B	B	B		A	A	A
Lactic acid	A	A	A	A	A	C	A	A	A	A	A	A	C		A	Z	Z	A		Z		A	A	A		A	A	
Lead acetate	C	A	A	A	A	A	A	A	A	A		A	A	A		A	Z	Z	A	A	A	A	B	B	C	A	A	A
Magnesium chloride	C	C	A	A	A	B	B	A	A	A	A	A	Z		A	A	A	A	Z	C		Z	C	Z		A	A	A
Magnesium nitrate	A		A	A	A	B	B	A	A	A	A	A		A		A	A	A		A	A	B	B	B		A	A	A
Magnesium sulfate	A		A	A	A	A	A	A	A	A	C	A	C		A	A	A	A	Z	A	A	A	A	A	C	A	A	A
Maleic acid	A		A	A	A	A	A	A	A		A	A	C	A	A	A	A	A	Z	A	A	A	A	A		A	A	A
Methanol	A	C	A	A	A	A	A	A	C	A	A	C			A	A	A	A	Z	A	B	B	A	B	C	A	A	A
Methyl chloride	A	C	A	A	A	A	A	A	A	A	A	Z		Z	A	Z	Z	Z	C		C	Z	Z	Z	Z	A	A	A

(Continued)

Table 10.2 (Continued)

	Brass	Bronze	Alloy 20	Hastelloy C	Monel	304 stainless steel	316 stainless steel	Titanium	Silicon iron	Tantalum	Copper	Aluminum	Carbon steel	Butyl rubber	Epoxy	Hypalon	Natural rubber	Neoprene	Nitrile rubber	Nylon	Phenolic	Polyethylene	Polypropylene	PVC	Silicone elastomer	TFE	Ceramic, alumina	Graphite
Methyl ethyl ketone	A	A	A	A	A	A	A	A	A	A	A	A	A	A	B	Z		Z		A	Z	A	B	Z		A	A	A
Methylene chloride	Z	Z	A	A	Z	A	A	A	A	A	Z	A			A	Z	Z	Z	Z	Z		Z	Z	Z	C	A	A	A
Napthalene	A	Z	A	A	A	A	A	A	A	A		Z			A	Z	A	Z	Z	A	A	A	B	Z	C	A	A	A
Nickel chloride	Z	C	A	A	A	A	A	A	A	A		Z	Z	A	A	A	A	A	Z	A	A	A	A	A		A	A	A
Nickel sulfate	Z	Z	A	A	Z	A	A	A	A			B	Z	A	A	A	B	Z	Z	Z		Z	A	A	Z	A	A	A
Nitric acid, 10%	Z	Z	A	A	Z	A	A	A	A	A	Z	B	Z	A	A	A	Z	Z	Z	Z		Z	A	A	Z	A	A	A
Nitric acid, 20%	Z	Z	A	A	Z	A	C	A	A	A	Z	B	Z	A	A	A	Z	Z	Z	Z		Z	B	A		A	A	N
Nitric acid, 50%	Z	Z	A	A	Z	A	C	A	A	A	Z	B	Z		A	A	Z	Z	Z	Z		Z	Z	A		A	A	A
Nitric acid, anhydrous	A	A	A	A	A	A	C	A	A	A	C	A	C		A	Z	Z	Z		Z	A	A	C	A		A	A	A
Oleic acid	A	A	A	A	A	A	B	A	A	A	C	C	Z		A	Z		C		C	A	A	A	A	Z	A	A	A
Oxalic acid	A	A	A	A	A	A	C	A	A	A	C	B	O		C	C	A	A	Z	A	A	A	A	A	A	A	A	A
Phenol	A	C	A	A	C	C	C	A	A	A	C	Z	C	A	A	Z	B	C		C	C	A	A	A	A	A	A	A
Phosphoric acid, 0–50%	C	C	A	A	A	C	C	B	A			Z	C	A	C	Z	C	Z	Z	A		A	A	A		A	A	A
Phosphoric acid, 50–100%	C	C	C	A	A	C	C	B	A	A	C	Z	C	A	C	Z	C	C		C	A	A	A	C		A	A	A
Potassium bicarbonate	A		A	A	A	A	A	A	A		A	A			A	A	B	A		A		A	A	A	A	A	A	A
Potassium bromide	A	A	A	A	A	A	A	A	A	Z	A	A			A	A	A	A	C	A	A	A	A	A	A	A	A	A
Potassium carbonate	Z		A	A	A	A	A	A	A	A	Z	Z	C	A	C	A	C	A	Z	C	A	B	A	A	A	A	A	A
Potassium chlorate	C	C	A	A	A	C	A	A	A	A	Z	Z		A	C	A	A	A	Z	C		A	A	A	A	A	A	A
Potassium chloride	Z	Z	C	A	A	C	C	A	A	A	Z	A	C	A	A	A	A	A	C	A		A	A	A	A	A	Z	A
Potassium cyanide	Z	Z	C	A	A	C	A	A	A	A	Z	Z	A		A	A	A	A	A	Z	A	A	A	A		A	A	A
Potassium dichromate	Z	Z	A	A	A	A	A	A	A	A	A	A		A	A	A	C	A	Z	A	A	A	A	A	A	A	A	A
Potassium hydroxide	N	A	A	A	A	A	C	A	A	A	A	Z		A	A	C	C	A	Z	C	A	B	A	A		A	Z	A
Potassium nitrate	A	C	A	A	A	A	A	A	A	A	A	A		A	A	A	A	A		A		A	A	A	A	A	A	A
Potassium permanganate	C	C	A	A	A	C	C	A	A	A	A	A	C		A	A	C	A		Z		A	A	A	A	A	A	A
Potassium sulfate	A	C	A	A	A	A	A	A	A	A	A	A		A	A	A	A	A	Z	C	A	A	B	B		A	A	A
Propyl alcohol	A	A	A	A	A	A	A	A	A	A	A	A		A	A	A		A		Z	A	A	A	A		A	A	A
Sodium acetate	A	C	A	A	A	A	A	A	A	A	A	A			A	A		A	Z	A	A	A	A	A		A	A	A
Sodium bicarbonate	Z	C	A	A	A	A	A	A	A	Z	A	A		Z	A	A	C	A	Z	C	A	A	B	A	Z	A	A	A
Sodium bisulfate	A	A	C	A	A	A	A	A	A	A	C	C		A	A	A	C	A		C	A	A	A	A	C	A	A	A
Sodium bisulfite	A		A	A	A	A	C	A	N	A	A	A	N		A	C	C	A	Z	C	A	A	A	A		A	A	A

(Continued)

Chemical	Graphite	Ceramic, alumina	TFE	Silicone elastomer	PVC	Polypropylene	Polyethylene	Phenolic	Nylon	Nitrile rubber	Neoprene	Natural rubber	Hypalon	Epoxy	Butyl rubber	Carbon steel	Aluminum	Copper	Tantalum	Silicon iron	Titanium	316 stainless steel	304 stainless steel	Monel	Hastelloy C	Alloy 20	Bronze	Brass
Sodium carbonate	A	A	A	N	A	A	A	A	A	Z	A	A	A	A	Z	A	N	A	N	A	A	A	A	A	A	A	A	A
Sodium chlorate	A	A	A	C	A	A	A	A	C		A	C	C	A	A					A	A	C	C		C	C		A
Sodium chloride	A	A	A		A	A	A	A	A	Z	A	A	A	A		C				A	A	C	C		C	C	Z	Z
Sodium cyanide	A	N	A	C	A	A	A	C	A		A		A	A		A	Z	Z	Z	Z	A	A	A	A	A	A	C	Z
Sodium hydroxide, 20%	A	N	A	A	A	A	A	C	A	Z	A	C	A	A	A	A	Z	C	Z	Z	A	A	A	A	A	A	C	Z
Sodium hydroxide, 50%	A	A	A	C	A	A	A	A	A	A	A	C	C	A	A	Z	Z	C	Z	Z	A	A	A	A	A	A	C	Z
Sodium hypochlorite	A	A	A	C	A	A	A		Z	A	C	C	A		A	A	Z	C	A	A	A	C	C	C	C	A	C	Z
Sodium nitrate	A	A	A	A	A	A	A	A	A	Z	A	C	A	A			A	A	A	A	A	A	A	A	A	A	A	C
Sodium silicate	A	A	A		A	A	A			Z	A	A	A		A		A	A	A	C	A	A	A	A	A	A	A	A
Sodium sulfate	A	A	A	C	A	A	A	A			A	C	A	A	A		A	A	A	A	A	A	A	C	A	A	Z	A
Sodium sulfide	A	A	A	C	A	A	A	A	A	Z	A	C	A	A	Z	C	Z	Z	Z	A	A	C	C	C	A	A	Z	Z
Stannic chloride	A	A	A	N	A	A	A			Z	A	C	A	A	Z	Z	Z	N	A	Z	A	Z	Z	B	C	B	C	Z
Stearic acid	A	A	A		A	B	C			Z	A	Z	C	A	A	Z	A	A	A	A	B	A	A	C	A	A	A	A
Sulfuric acid, 0–10%	A	A	A		A	A	C				A	Z	A	A	A	Z	Z	Z	A	Z	C	Z	Z	C	A	A	A	Z
Sulfuric acid, 10–75%	A	A	A	Z	A	A	C	C	Z	Z	Z	Z	A	A	A	Z	Z	Z	A	Z	Z	Z	Z	C	A	A	A	Z
Sulfuric acid, 75–100%	A	A	A	Z	B	B	A	C	Z	Z	Z	Z	C	A	A	Z	Z	Z	A	A	C	Z	Z	C	C	A	A	Z
Tannic acid	A	A	A		A	A	A	C	Z	Z	A	A	A	A	Z		Z	A	A	A	A	C	C	A	A	A	A	A
Tartaric acid	A	A	A		A	A	C		Z		A	A	A	A			A	C	A		A	A	A	C	A	A	C	C
Tetrahydrofurane	A	A	A		Z	B	Z									A						A	A		A	A		
Toluene	A	A	A	Z	Z	B	Z	A	A	Z	Z	C	Z	C	Z	A	A	A	A	A	A	A	A	A	A	A	Z	A
Trichloroethylene, dry	A	A	A	Z	Z	B	Z	C	Z	Z	Z	Z	Z	A	Z	A	A	A	A	A	A	A	A	A	A	A	Z	A
Turpentine	A	A	A	Z	Z	B	Z	A	Z	C	Z	Z	Z	B	Z	C	A	C	A	A	A	A	A		A	A	C	C
Urea	A	A	A		A	A	A	A	A		A		A			N	A					A	A		A	A		
Xylene	A	A	A		Z	Z	Z				Z		Z	C			N	N	A	A	A	A	A	A	A	A	N	N
Zinc chloride	A	A	A	A	A	A	A	A	Z	Z	A	C	A	A	A	C		C	A	A	A	A	A	A	A	A	Z	Z
Zinc sulfate	A	A	A	A	A	A	A	A	C	Z	A	C	A	A	A							A	A		A	A	A	C

A, acceptable; B, acceptable up to 80°F; C, caution, use under limited conditions; N, not recommended; blank space, effect unknown. (Reproduced from Sandler, H.J. and Luckiewicz, E.T. (1987) Practical Process Engineering: A Working Approach to Plant Design).

I usually favor the Langelier index (LSI) for historical reasons but for carbonate buffered systems the Ryznar/Carrier Stability Index (RSI) is supposedly more empirically based.

The Larson-Skold index predicts corrosion of mild steel and, since it considers sulfate and chloride as well as bicarbonate, is commonly used to predict corrosivity of once-through cooling seawater.

The Oddo-Tomson index allows for water, gas, and oil phases, and the effect of pressure on CO_2 saturation, to satisfy oil and gas industry needs.

MECHANICAL EQUIPMENT

Though the average new graduate knows very little about pipe specification and so on, I will focus in the section on qualitative information on the most important items of mechanical equipment which are largely untaught in chemical engineering degrees. These are in my opinion fluid transport, flow control and heat exchanger selection.

Pumps/blowers/compressors/fans

Fluid moving equipment generally comes in two main varieties—rotodynamic or positive displacement. Each of these comes in many subtypes, but beginners have often not been taught the crucial differences between the two broad types.

Later, I provide a table giving you some ideas on how to choose between the commonest types of liquid and gas moving equipment, but first let's see how we choose between the broad varieties of pumps (Table 10.3).

Table 10.3 Pump selection (general)

	Rotodynamic	Positive displacement
Head	Low—up to a few bar	High—hundreds of bar
Solids tolerance	Low without efficiency losses	Very high for most types
Viscosity	Low viscosity fluids only	Low and high viscosity fluids
Sealing arrangements	Rotating shaft seal required	No rotating shaft seal
Volumetric capacity	High	Lower
Turndown	Limited	Excellent
Precision	Low—discharge proportional to backpressure	Excellent—discharge largely independent of backpressure
Pulsation	Smooth output	Pulsating output
Resistance to reverse flow	Very low	Very high
Reaction to closed valve downstream	No damage to pump	Pump damage likely

This is why we tend to use positive displacement pumps for metering duties, and centrifugal (rotodynamic) pumps for moving large volumes of flow at relatively low pressures. More detailed choices can be given in Table 10.4.

Table 10.4 Pump selection (detailed)

	Relative price	Environmental/safety/operability concerns	Robustness	Shear	Maximum differential pressure	Capacity range	Solids handling capacity	Efficiency[a]	Seal in-out	Fluids handled[b]	Self-priming?
Rotodynamic											
Radial flow	M	Cavitation	H	H	M/H	L–VH	M	H	M	Low viscosity/aggressiveness	N
Mixed flow	M	Cavitation	H	H	M	M–VH	M	H	M	Low viscosity/aggressiveness	N
Axial flow	M	Cavitation	H	H	M	M–VH	M	H	M	Low viscosity/aggressiveness	N
Archimedean screw	H	Release of dissolved gases	VH	H	L	M–VH	H	L	N/A	Low viscosity/aggressiveness	N
Positive displacement											
Diaphragm	M	Overpressure on blockage	VH	L	M/H	VL–M	H	L	H	Low/high viscosity/aggressiveness	Y
Piston diaphragm	H	Overpressure on blockage	M	L	VH	VL–M	L/M	M	H	Low/high viscosity/aggressiveness	Y
Ram	H	Overpressure on blockage	H	M	VH	M–H	H	M	H	Low/high viscosity/aggressiveness	Y
Progressing cavity	M	Overpressure on blockage	M	VL	H	M–H	H	H	H	Low/high viscosity/aggressiveness	Y
Peristaltic	H	Overpressure on blockage	L	VL	M	VL–M	H	M/H	M	Low/high viscosity/aggressiveness	Y
Gear	L	Overpressure on blockage	L	M	VH	VL–L	L	L	H	Low/high viscosity. Low aggressiveness	Y
Screw	L	Overpressure on blockage	L	M	VH	L–M	L	L	H	Low/high viscosity. Low aggressiveness	Y
Other											
Air lift	L		VH	VL	VL	L–M	H	L	L	Low viscosity, low/high aggressiveness	N
Eductor	M	Blockage of eductor	M	M	VL	L–H	M	H	M	Low viscosity, low/high aggressiveness	Y

[a]Centrifugal pump efficiency reduces as viscosity increases, but PD pump efficiency increases. Centrifugal pump efficiency is more to do with impeller type than anything else, impeller type is determined by process conditions such as any solids handling requirement.
[b]Aggressiveness is related to presence of abrasive particles, undissolved gases, or unfavorable LSI.

Valves

We can think of valves in a number of broad categories. It can incidentally be helpful, when first constructing a Piping and Instrumentation Diagram (P&ID) of a process, to add them using these categories in turn.

The most common types on a process plant are usually the valves which allow every item of equipment to be capable of isolation from the rest of the plant for maintenance purposes. The (usually manually operated) valves we use to do this can be called isolation valves. They are usually set to either their fully open or fully closed position, rather than being used for flowrate control. These are usually tagged on a P&ID as MV—manual valve (though some think nowadays that this clutters the P&ID, and do not tag MVs).

A variant on isolation which can be useful (but carries a significant risk) is a bypass. A bypass valve allows a unit operation to be bypassed. If a complete or partial bypassing of a unit operation carries safety or performance implications (as it usually does), the designer needs to think about how to protect the plant from accidental or deliberate bypassing of a unit operation by operators. Valves can be locked out, or a section of pipe known as a spool piece can be left out and kept under lock and key to make sure that bypassing is only done deliberately and under management control.

Bypassing of actuated control valves via a manual control valve is more akin to a manual standby unit, and is far less risky from a process point of view than bypassing of unit operations.

There are a smaller number of valves on a plant which are used to control flow (control valves). They may be used for on/off control, or they may be used to modulate flow by intermediate degrees of opening. The setting of these valves may be manual, or more frequently nowadays they are moved by means of motors known as actuators, controlled by computer. These may be tagged on a P&ID as AV—actuated valve, or less desirably FCV—flow control valve. They should not be tagged as MV—motorized valve, to avoid confusion with the last category.

Then there are the self-operating safety valves, including nonreturn valves, pressure relief valves, and pressure sustaining valves.

My final category is externally operated safety valves such as emergency shutdown valves.

The preceding list is of valve duties rather than valve types. In order to choose an appropriate type of valve for a duty we need to know the characteristics of the various types of valves commercially available. Table 10.5 gives this information to allow a choice to be made.

There is also a handy table in Practical Process Engineering (see Further Reading) which I reproduce in Table 10.6.

Table 10.5 Valve selection

	Relative price	Robustness	Sizes (mm NB)	Materials of construction	Fluids handled	Solids handling ability	Seal in-out	Seal up-downstream	Controllability	Actuator type	Pressure rating[a]	Manufacturer[b]
Isolating												
Butterfly/disc	L	M	50–500	Plastics, cast iron		M	M	M	M/H	1/4 turn cylinder, electric motor	M/H	Bray
Globe	L	M	50–600	Brass		M	L/M	M	M/H	Multiturn electric motor, linear cylinder	M	Vela
Ball	L	H	5–1,200	Plastics		M/H	M	H	M/H		H	George Fischer
Diaphragm	H	H	15–500	Plastics, stainless steel, aluminum		H	H	H	L/M		M	Saunders
Gate	M	M	50–1,250	Cast iron		H	L/M	H	M/H		H	GWC
Control												
Needle	H	M	6–25	Stainless steel		L	M	H	H		H	Hoke
Globe	L	M	50–600	Brass		M	L/M	M	M/H		M	Velan
Plug	H	M	15–1,800	Cast, ductile iron, bronze, aluminum, carbon steel, stainless steel, alloy 20, and Monel	Clean and dirty liquids and gases, sludge, and slurries	H	M	M/H	H	Lever, hand wheel, chain wheel, cylinder, electric motor	M	De Zurik
V-Notch ball	L	H	6–150	Plastics, stainless steel		M	M	H	M/H		M/H	George Fischer
Eccentric disc	H	M	50–1,200			M	M	H	M/H		M	Neles
Other												
Emergency shutdown	H	H	50–400			M	M	H	H		M/H	Becker
Pressure relief	H	M	6–900	Brass, cast iron, stainless steel		L	L/M	L/M	NA		M	Farris
Swing check	M	H	50–900	Cast iron		H	H	M	NA		M	Velan
Spring check	M	M	6–600	Cast iron		M	M/H	—	NA		M	Flowserve
Pressure sustaining	H	M	6–350	Plastics		L			M/H		M	Bermad
3-Way ball	M	M	12.5–150	Plastic, cast iron		M/H	M	H	M/H		M	George Fischer
3-Way plug	M/H	H	80–400	Cast iron, Ni-resist, aluminum, carbon steel, 316 stainless steel	Clean and dirty, viscous and corrosive liquids, sludge, abrasive and fibrous slurries, and clean and dirty corrosive gases	H	M	M/H	M/H	Lever, hand wheel, chain wheel, cylinder, electric motor	M/H	DeZurik

[a] There are ISO/API classes of valve by pressure/temperature ratings.
[b] Not a recommendation, just a route to more information.

Table 10.6 Primary usages for various common valve types

Type	Neutral (water, oil, etc.) — On-off	Neutral (water, oil, etc.) — Reg	Corrosive (alkaline, acid, etc.) — On-off	Corrosive (alkaline, acid, etc.) — Reg	Hygienic (beverages, foods, drugs) — On-off	Hygienic (beverages, foods, drugs) — Reg	Slurry — On-off	Slurry — Reg	Fibrous suspensions — On-off and Reg	Neutral (air, steam, N_2, etc.) — On-off	Neutral (air, steam, N_2, etc.) — Reg	Corrosive (acid vapors, chloride, etc.) — On-off	Corrosive (acid vapors, chloride, etc.) — Reg	Vacuum — On-off	Vacuum — Reg	Abrasive powder (silica, etc.) — On-off and Reg	Lubricating powder (graphite, talc, etc.) — On-off and Reg
	Liquid flows									**Gaseous flows**						**Solid flows**	
Gate	●		●					●	●	●	●			●			●
Globe	●	●								●	●		●	●	●		
Ball	●			●			●			●		●		●			
Plug	●		●				●			●		●					
Butterfly	●	●	●	●	●	●	●	●		●	●	●	●	●	●		
Diaphragm		●	●	●	●	●	●	●	●		●	●	●				
Flush-bottom			●				●	●	●								
Squeeze						●	●	●	●							●	●
Pinch						●	●	●	●							●	●
Needle		●									●		●				
Slide gate																●	
Spiral sock																●	●

(Reproduced from Sandler, H.J. and Luckiewicz, E.T. (1987) Practical Process Engineering: A Working Approach to Plant Design).

Heat exchangers

All new graduate chemical engineers can perform a "shortcut design" for a heat exchanger, but few really know of more than one type—the shell and tube exchanger which the shortcut method applies to.

The kind of detailed design taught in universities is usually a matter for equipment suppliers rather than process plant designers. There are, however, quite a number of types of heat exchangers, and selection between them is part of the process plant designer's job.

Table 10.7 is intended to help the beginner to select a suitable heat exchanger type or types.

Table 10.7 Heat exchanger selection

	Relative initial cost	Robustness	Sizes (kW)	Materials of construction	Fluids handled	Solids handling ability	Fouling resistance	Maintainability (cleaning)	Sealing	Max design pressure	Max design temperature	Hygienic operation?	Efficiency	Footprint	Required temperature of approach	Pressure drop
Shell and tube	M	VH	H	Flexible, depends on corrosion study	Liquid, gas or two-phase	M; To select appropriate pitch and type	L	M; To select appropriate pitch and type depending on service	H	H	H	N	M	H	H	L
Spiral	L (for low duty)	VH	L	Flexible, depends on corrosion study	Liquid, gas or two-phase	VH	L	H	H	H	H	Y	L	VH	H	L
Double pipe	L (for low duty)	VH	L	Flexible, depends on corrosion study	Liquid, gas or two-phase	H	L	H	H	H	H	Y	L	VH	H	L
Printed circuit heat exchanger	H	VH	H	Flexible, depends on corrosion study	Liquid, gas or two-phase	L; Depends on particle size and flow passage size; Prior filtration may be required. Generally only for clean service	H	VL (chemical cleaning only; mechanical cleaning is not possible)	VH	VH	VH	N	VH	VL	VL	VH
Plate and shell	VL	H (plate failure)	H	Ensure fluid handled compatible with gasket	Normally Liquid, rarely gas or two-phase	L; Prone to blockage	H	VH	M	M	H	Y	H	L	VL (as low as 1°C)	H
Plate and frame	VL	L (plate failure, gasket failure)	H	Ensure fluid handled compatible with gasket	Normally liquid, rarely gas or two-phase	L; Prone to blockage	H	VH	L	L (up to 25 bar g)	L (Design temperature limited by gasket material)	Y	H	L	VL (as low as 1°C)	H

ELECTRICAL AND CONTROL EQUIPMENT

Chemical engineering students do not really think about power and control for the systems they "design," other than in the most abstract mathematical terms. An understanding of the needs of electrical and control engineers is, however, crucial to competent design.

Here are the most important items for consideration:

Motor control centers

When I first started designing plants, I did not know what a Motor Control Center (MCC) was, why it was needed, and what it contained. (This is a standard feature of Chemical Engineering degrees: we don't get taught about the most basic needs of the other disciplines we will be interacting with.) This became a bit of a problem when I almost won a job for my employer in which I hadn't thought it necessary to include an MCC. To save you the same embarrassment, allow me to explain.

Electrical motors or "drives" require maybe six times their running power to start them up. Rather than uprating all the power cabling and so on to this starting current, motor starters are used to send a pulse of power to get the drive spinning. They also contain overload protection and so on.

Direct on Line (DOL) starters are very cheap, but they simply apply the full line current to the motor all at once in a way which usually limits their use to drives rated at less than 11 kW.

Star Delta starters are more expensive. They apply current to the motor in two configurations in a way that reduces starting torque by a factor of three. These are probably required above an 11 kW drive rating.

Soft Starters are the most expensive type. They control voltage during drive start-up in a way which avoids the torque and current peaks associated with DOL and Star Delta starters, and have some of the sophisticated control functions of the variable speed drive (VSD).

Inverters or VSDs are perhaps a little more expensive than Star Delta starters, but they have a lot more flexibility. They allow very sophisticated patterns of power ramping to be applied to the drive on start-up, as well as allowing variable frequency to be supplied in a way which allows drive rotational speed to be controlled. They always have a microprocessor on-board nowadays so that multiple, quite sophisticated control loops and interlocks can be run directly through them.

These drives are usually collected in a big box called an MCC, which will usually contain internal subdivisions housing starters or groups of starters. There are some other common features, as shown in Figure 10.1:

Figure 10.1 Photograph of an MCC unit, showing incomer, starters, marshalling cubicle, and PLC/UPS sections. *Copyright image reproduced courtesy of the Process Engineering Group, SLR Consulting Ltd.*

The image shows a "Form 4" panel, with an intermediate degree of separation between controls for different aspects of a process. There are four forms of panel specified in IEC 60439-1: 1999, Annex D and BS EN 60439-1: 1999. There are lettered subtypes, but broadly:

Form 1 has no separation, and is often referred to as a wardrobe type. Failure of one component in a Form 1 panel can damage other components, and a single failure will take the whole process offline.

Form 2 separates the bus bars (big copper conductors which carry common main power throughout the panel) from other components. Not much better than Form 1.

Form 3 separates the bus bars from other components, and all components from each other. This is the minimum specification if you would like sections of the plant to be capable of running while one of them is offline.

Form 4 is as Form 3, but also separates terminals for external conductors from each other.

There is a different system in the United States (described in standard UL 508A) which takes a different approach, but addresses the same issues. In either case the process designer will be required to specify broadly which kind of panel they want, though clients may specify a minimum separation between control switchgear.

Panels will also have a specified degree of ingress protection (IP55 is usually the minimum standard).

Consideration will also need to be given to direction of cable entry, which can be from the top, bottom, or a combination of both. Bottom entry requires the panel to sit on a channel in the floor, so there are civil engineering implications.

There is great deal more to this, but this is the minimum level of knowledge required of a process designer to integrate their MCC design.

Cabling

Another thing which generally receives no attention in chemical engineering degrees is cabling. The absolute minimum information you need to know about cables is as follows:

Every instrument needs incoming power and outgoing signal cabling. Every electrical drive (motor) on the plant needs incoming power cabling. Every MCC needs incoming power cabling and outgoing power and signals cabling.

Power cable size is calculated by electrical engineers in a similar way to pipe size. The more current a cable carries, the thicker it has to be to avoid overheating. A complicating factor is that cables can have a variable number of wires or "cores" inside them. Thus we can have a thick cable with a single set of large cores feeding a single large drive, or a similarly thick cable with multiple smaller cores to feed a number of drives.

Instrument cabling needs to be arranged so as to be unaffected by electromagnetic fields from the power cabling. This is usually achieved by some combination of physical separation and shielding. 1 m of separation will usually do it, but your electrical engineer will advise.

The basic kind of power cable is the relatively inflexible SWA (Steel Wire Armored) PVC-insulated type. There is also a more flexible, unarmored, and waterproof kind used to connect submersible pumps. Instrument cabling also comes in a number of types, and is a lot smaller than power cabling as it is only carrying 5—240 V, and no real power.

Cables have a minimum bend radius. The thicker the cable, the bigger this is. There needs to be space in the design to accommodate this. It is approximately equal to the radius of a commercially available long radius bend in a pipe large enough to carry the cable.

All kinds of cable are carried underground through ducts, which are nowadays plastic pipes. These may be cast into concrete slabs, unlike pipes carrying fluids.

We usually specify a few more such ducts than we need in a design to allow for future expansion. Cables are carried overground on (usually elevated) cable trays. Power and signals cabling should ideally be in separate ducts and cable trays.

Instrumentation

There are many specialized process instruments, but the four commonly measured parameters are pressure, flow, temperature, and level, and process designers need to be able to choose between the most common types of instrumentation used for measuring these things. Table 10.8 should help:

Table 10.8 Instrumentation selection

	Relative price	Robustness	Contact with process fluid?	Solids handling ability
Pressure instruments				
Bourdon	L	L	Y	M
Capacitance	M	H	Y	H
Resistance	L	M	Y	H
Piezoelectric	M	H	N	H
Optical	H	M	N	M
Flow instruments				
Variable area	L	L	Y	L/M
Mechanical	L	L	Y	
Pressure	M	M	Y	L/M
Electronic	M/H	H	N	H
Radiation	H	H	N	H
Vortex	H	H	N	H
Doppler	M	H	N	VH
Temperature instruments				
Thermocouple	L	H	Y	H
Bimetal	L	M	Y	M/H
Resistance	L	H	Y	H
Level				
Sonar/Radar	M	M	N	H
Float	L	M	Y	H
Pressure	M	M	Y	M/H

Control system

A number of control systems are used in process plant design. Selection between them for whole-system or local design may be as much a matter of client preference, industry familiarity, and designer preference as inherent characteristics of the system type.

Local controllers

Once upon a time, control loops were operated by mechanical, electromechanical, pneumatic, or electrically operated boxes which were mounted locally to the thing being controlled.

By the time I was at university, these were most commonly solid state electronic PID (Proportional, Integral, Differential) controllers (Figure 10.2).

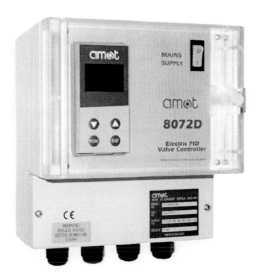

Figure 10.2 Wall-mounted PID controller. *Copyright image reproduced courtesy of Amot.*

We still use the odd dedicated field-mounted controller (for pH control, for example) but process plant designers never do the thing we were taught to do in university—writing algorithms for these boxes. They are products, whose manufacturers have done this job for us. Their limited configurability also makes these controllers the most robust solution. Writing software is best left to experts.

The oil and gas industry still use the modern equivalent of P&ID controllers, and a DCS system to facilitate their aftermarket optimization and tuning of plants, but this is not much to do with process plant design as I define it here.

Programmable logic controllers

In my industry, programmable logic controllers (PLCs) are commonly used for whole system control. PLCs are a kind of computer which is custom built out of components to suit a particular duty.

There is a range of central processing units of increasing power available, and rack-mounted cards are added to this to provide a suitable number of input and output channels.

Direct interface with a PLC is via a human—machine interface (HMI) or PC program. These can vary in appearance from the program's green-screen and ladder logic to sophisticated simulation interfaces of supervisory control and data acquisition (SCADA) systems.

It should be noted that, while PLCs themselves cannot be directly infected with computer malware, the intelligence community produced a worm (Stuxnet) which can attack PLCs via their SCADA connected PCs (thought to have been written to target the Iranian nuclear program). Stuxnet was purely destructive, but a more recent virus called Duqu is a keystroke-logging spyware program.

Any systems which include a PC may be compromised, especially as PCs are almost always connected to the internet nowadays, and site communications and signals are increasingly connected via Wi-Fi rather than hard wired.

PC

Supervisory control and data acquisition

SCADA systems run on a PC. They can receive signals from one or more PLCs, or from remote telemetry outstations (RTUs) which convert 4—20 mA signals from field instruments into digital data.

Such signals may be carried by local or wide area networks using internet protocols, by telephone lines or satellite signals.

The SCADA system has a HMI—usually in the form of simulation screens which look rather like animated PFDs, alarm handling screens, "trends" screens which allow variation in parameters to be seen as a graph against time, and input screens which allow process parameters to be changed (by authorized users).

DCS

DCS used to be more different from SCADA than it is nowadays. Historically SCADA used dumb field instrumentation, and had a centralized system brain, whereas DCS has a lot more control out in the field.

As it is getting increasingly difficult to buy dumb instrumentation, or even dumb motor starters, the distinction is not as sharp as it was, but DCS is more likely to involve field-mounted controllers. These are less likely to be simple PID controllers, but will be capable of more sophisticated control.

The oil and gas industry makes extensive use of DCS, which is well suited to their culture of radical postpurchase control system tuning.

Aftermarket systems: supervisory computer, etc.

In the oil and gas industry, there is a tendency to install postpurchase a supervisory control system running advanced control software, such as MPC (multivariable predictive control) and RTO (real-time optimization). To quote Myke King:

> MPC is installed on pretty much every oil refinery and petrochemical site, but it's not something that needs a lot of attention at the process design stage.
>
> It and RTO (if justified) would typically be engineered well after process commissioning—in some cases 30 years after commissioning! There would be other PCs connected to the DCS for process data collection—typically based on a real-time database (such as OSI's PI, Honeywell's PHD etc.). They would have links to other process management systems—such as LIMS (laboratory information management system).

FURTHER READING

Branan, C., 2012. Rules of Thumb for Chemical Engineers. Elsevier.
Couper, J.R., Penney, W., 2012. Chemical Process Equipment—Selection and Design. Elsevier.

CHAPTER 11

How to Design Unit Operations

INTRODUCTION

Process plant designers very rarely carry out detailed design of unit operations, as they do not intend to offer the process guarantees for the units. Detailed design is a job for those guaranteeing the performance of the unit operation.

The plant designer's job is to get their design correct enough such that the unit operation will work reliably across the design envelope, as the process guarantee will limit its validity to that design envelope. We also frequently check that there are no misunderstandings in what vendors have offered or any design details incompatible with the broader design.

We will also usually make sure that the equipment weight, size, power, and other utilities requirements are in line with our expectations. If they are not, we may need to consider modifying the whole plant design to suit, or reject the item of equipment as it is less favored when these knock-on effects are considered.

MATCHING DESIGN RIGOR WITH STAGE OF DESIGN

We do not wish to spend any more time on design at each stage than is necessary to progress the overall design. Rule of thumb design is therefore the norm for process plant designers.

We do, however, use more onerous but accurate heuristics as it grows increasingly likely that some will actually build the plant.

RULE OF THUMB DESIGN

If you are working as a process designer, there will mostly (but not always) be a design manual which will give the relevant rules of thumb for designing the items you are being asked to design which encapsulates the company's experience in the area.

Failing that, there will be a more experienced engineer in your department or at least in your company who knows the rules of thumb. If there is neither a manual nor such a person, this is a bad sign. You aren't going to learn much in this company.

If there is such a person, they may not be willing to share their knowledge. This isn't a great sign, but sooner or later the company will have to give you support if they want you to do a competent job.

The more experienced engineer is often the world's leading expert on the particular job you have to do, as they don't just know a way to design the plant, but they know how to design the plant so that your company can build it. So be nice to him or her, as they can teach you more than university ever did.

But what if you don't have a manual or a Yoda? As well as Perry's Handbook, there are books which contain general rules of thumb—I give a couple of good ones in the reading list at the end of the chapter. These are not going to be as good as the experienced engineer, but they may well be more useful than attempting to use some of the theoretical approaches they taught you at university.

There are some recent books offering "rules of thumb" generated by modeling and simulation programs. Don't use these. Proper rules of thumb come from experience with multiple full-scale real-world plants, not first principles computer programs. First principles design doesn't work.

APPROACHES TO DESIGN OF UNIT OPERATIONS

First principles design

Don't ever do this in normal professional process plant design practice. Even if you had enough data to allow you to design a unit operation from first principles, it would be at best a prototype, and your employer would be the one offering the process guarantee by making it part of the plant they were guaranteeing.

If you are designing unit operations for a living, you are not really a process plant designer, so I will not cover this issue in much detail in this book about process plant design. You probably have colleagues who have the necessary know-how or, more likely still, a spreadsheet or program written by someone else which encapsulates the know-how.

If you are asked to author such a spreadsheet or program, make sure it does reflect the experience held in the company, that the program is without bugs, and verify its accuracy across its range of operation by full-scale real-world experiments. It will have fewer bugs if you write it in Excel rather than compiled code.

Design by simulation program

Despite all that I have said about the use of simulation at whole-plant level, unit operations are less complex than whole plants, and some suppliers now offer you simulation program blocks which they have verified and tuned to match their real equipment (though not validated for your application).

Companies which offer plants which are made of blocks of a limited number of unit operations running on basically invariant feedstocks (as when, e.g., producing nitrogen from air) can produce model blocks which encapsulate much empirical data.

In this way the most normal of normal process design activity can become rather similar to building a Lego model. I am not, however, sure that we can call this activity process plant design. It seems to me more like equipment design, where a whole plant may be specified as a collection of standard equipment.

Design from manufacturers' literature

Since detailed design involves putting together unit operations you can actually buy, manufacturers' catalogues are a useful tool for selecting the unit operations which we put into our plant designs.

Updated catalogues also frequently include new items of equipment we might not have considered if we had not read the catalogue.

Back when I only designed plants for a living, reading through the pile of supplier catalogues which had accumulated while I was engrossed in designing and pricing the last plant was a very useful way to spend the time waiting to be allocated my next job.

Nowadays these catalogues are more likely to be found on websites rather than in hardcopy, and this is very handy in an academic setting, allowing us to bring realism to our students' designs without bothering manufacturers with enquiries from students who are not actually going to purchase anything.

One thing which we find hard to recreate in the academic setting is interactions with technical sales staff. Their detailed knowledge of their products and their capabilities and limitations can allow plant designers to see new ways to design ourselves ahead.

Civil engineers are, by the way, convinced that picking "gubbins" from catalogues is all process engineers do for a living. This is a point they may make to you when you tell them they can't have the data they need to proceed with their design for a couple of weeks. (The standard riposte to this is to tell them that whatever you say, they will in any case specify the concrete wall they are going to design to be 150 mm of concrete containing two sets of #10 M rebar.)

SOURCES OF DESIGN DATA

For the professional designer, the strength of sources of design methodology information is generally as follows (in descending order):

Close to full-scale pilot plant trial
Thoroughly developed and validated tailored modeling program, directly based on many full-scale installations of exactly the type and size of plant proposed
Supplier Information
Robust rule of thumb direct from Chartered Engineer with lots of relevant experience
Rules of thumb from a book by Chartered Engineer
Direct from Chartered Chemical Engineer with little relevant experience
First principles
Guessing by beginner
Less than thoroughly debugged and validated simulation and modeling program output

As far as sources of data required inputting to design methodologies is concerned, strengths are generally as follows for feedstock and product qualities and quantities:

Statistically significant ranges of values for the exact type and scale of plant envisaged
Statistically significant ranges of values for the type of plant envisaged
Ranges of values for the exact type of plant envisaged, falling short of statistical significance
Ranges of values for the broad type of plant envisaged, with or without statistical significance
Guessing by beginner

As far as thermodynamic and other physical and chemical data is concerned, the test of validity should be whether the data is intended by those generating it to be valid over the range of physical conditions likely to be encountered under all reasonably foreseeable plant operating conditions.

SCALE-UP AND SCALE-OUT

In theory, there is no difference between theory and practice. But in practice, there is.

Yogi Berra

Theoretical types and scientific researchers fondly imagine that if a reaction works in a conical flask, getting it to work in a 100 m^3 reactor is a pretty trivial thing. If this were true, there would be no chemical engineers. There is a lot more to chemical engineering than chemistry.

There are two basic approaches to making a bigger plant. We can have a lot of parallel streams of plants which we know to work at the given scale (scale-out), or we can have a smaller number of larger but at least slightly experimental plants (scale-up).

What we need to know as engineers to carry out a scale-up exercise is the critical variable or dimension. This variable is the thing we need to keep constant (or vary in a predictable way) in order to get the process to work at the larger scale.

We might need to maintain the length, area, or volume of a process stage, or it might be more complex, such as a number of theoretical plates for a distillation column.

It should be noted that the key variable might change at different stages of scale-up as the balance of effects varies. Scale-up by a factor of more than about 10 from even a good pilot plant study should make us quite nervous as designers.

We might also need to ask the chemists to go back and make the reaction work with less hazardous reactants or solvents, or in some other way restrict their freedom to get them to offer a process which can be made to work in an economically viable plant.

There is a lot to this subject, but the main thing to grasp is that something working in a 250 ml flask is no guarantee at all that it will be cost-effective, safe, or robust in a 25 l vessel, let alone at full scale.

NEGLECTED UNIT OPERATIONS: SEPARATION PROCESSES

Many students leave university with knowledge of only 6—10 separation processes, usually those most important in oil refining, and the knowledge they have of these processes is mostly at best engineering science.

Processes based in chemistry are overrepresented in university courses, as are liquid/liquid separations. The consequent lack of knowledge of options impoverishes the designer's imagination, and a lack of understanding of engineering design practice prevents practical use of technologies.

Table 11.1 provides an overview of the most important separation processes, arranged by phases separated. However, it is not exhaustive, does not go beyond separations of two components, and does not mention the plethora of subtypes of the technologies listed.

There is a book which fills this gap in the knowledge of professional plant designers, which is not much used in academia: Couper (see "Further Reading" for details).

Table 11.1 Separation processes

	Relative price	Process robustness	Safety concerns	Separation principle[a]	Product recovery per stage	Contaminant removal per stage
Gas/Gas						
	H	M	Rapidly rotating equipment	Sedimentation	M	M
Adsorption	M	M		Surface interaction		VH
Distillation	H	L/M	Flammable/toxic vapors, heat	Differential vaporization/ condensation	L/M	M/H
Filtration	L	H		Physical exclusion of oversize particles	H	H
Gas/Liquid						
Distillation	H	L/M	Flammable/toxic vapors, heat	Differential vaporization/ condensation	L/M	M/H
Adsorption	M	M		Surface interaction	H	VH
Stripping	L	H	Flammable/toxic vapors/liquids	Mass transfer from liquid to vapor phase	M	M
Filtration	L	H		Physical exclusion of oversize particles	H	H
Gas/Solid						
Adsorption	M	M		Surface interaction	H	VH
Freeze–drying	H	M	Low temperatures	Sublimation	H	VH
Filtration	L	H		Physical exclusion of oversize particles	H	H
Elutriation	M	M		Sedimentation	M/H	M/H
Cyclones	M	H		Sedimentation	M/H	M/H
Drying	M	H	Heat	Evaporation, usually heat assisted	VH	VH

Liquid/Liquid

Decantation	L	H		Sedimentation	M/H	H
Extraction	M	M		Mass transfer from one liquid phase to another	M	H
Membranes	M	L/M		Physical exclusion of oversize droplets	H	H
Chromatography	VH	L/M		Differential affinity	VH	VH

Liquid/Solid

Crystallization	M	M		Mass transfer from solution to crystal	M/H	H
Coagulation	L	M/H		Destabilization of a colloid	M	M
Precipitation	L	H		Exceeding solubility limit	M/H	M
Flotation	M	M/H		Sedimentation effected by reducing solid or liquid density with gas bubbles	H	H
Ion exchange	H	M		Differential affinity	VH	VH
Electrolysis	H	H	Electricity	Electrochemistry	H	H
Centrifugation	M	M	Rapidly rotating equipment	Sedimentation	M/H	H
Hydrocyclone	M	M		Sedimentation	M/H	M/H
Magnetic separation	H	M			VH	VH
Extraction/ Leaching	M	H		Mass transfer from solid to liquid phase	M/H	M/H
Chromatography	VH	L/M		Differential affinity	VH	VH
Membranes	H	M		Physical exclusion of oversize particles for coarser membranes, diffusion for RO	VH	H
Electrophoresis	VH	L/M		Electrochemistry plus drag effects	VH	VH

(Continued)

Table 11.1 (Continued)

	Relative price	Process robustness	Safety concerns	Separation principle[a]	Product recovery per stage	Contaminant removal per stage
Solid/Solid						
Classification	M	M/H		Sedimentation/ adhesion/electrostatic	M/H	H
Sublimation	M	H		Sublimation	VH	VH
Magnetic separation	H	M/H		Differential magnetic attraction	VH	H

Real-world separators differ from mathematical/theoretical ones in ways which usually make any first principles design unworkable.
[a]These principles are major influencers of the separation process, but they are at best tools for analysis and partial understanding.

FURTHER READING

Coulson, J.M., Richardson, J.F., 1995. Chemical Engineering: Fluid Flow, Heat Transfer and Mass Transfer. Butterworth-Heinemann, London.

Couper, J.R., Penney, W., 2012. Chemical Process Equipment—Selection and Design. Elsevier, Amsterdam.

Green, D.W., Perry, R.H., 2007. Perry's Chemical Engineers' Handbook. McGraw-Hill, New York, NY.

Hall, S., 2012. Rules of Thumb for Chemical Engineers. Elsevier, Oxford.

Sinnot, R.K., Towler, G., 2005. Chemical Engineering Design, Vol. 6. Butterworth-Heinemann, London.

Woods, D.R., 2007. Rules of Thumb in Engineering Practice. Wiley-VCH, Weinheim.

CHAPTER 12

How to Cost a Design

INTRODUCTION

Engineering ... to define rudely but not inaptly, is the art of doing that well with one dollar, which any bungler can do with two after a fashion.

Arthur Mellen Wellington

Engineering is a commercial activity. Sufficient effort is put into pricing at each stage of design to allow a rational commercial decision to be made as to whether to proceed to the next stage, but ideally no more. Costing itself has costs.

MATCHING DESIGN RIGOR WITH STAGE OF DESIGN

Conceptual design is sufficient for what contractors would call a budget estimate of costs. If you get a real budget estimate from a contractor, it will probably be accurate to around ±30%, as they have lots of data from equipment suppliers and genuine knowledge of just what it costs to engineer and build plants.

Beginners without this information and experience can produce estimates out by several hundred percent (almost always underestimates). They tend to leave out everything other than the very core of the production process, have unrealistic ideas of the cost of engineering and construction, no knowledge of the cost of engineering by other disciplines, and so on. Many of my students also seem to be willing to forgo profit, which is the whole point of engineering. They certainly frequently forget to add it to their estimates.

Beginners tend to use exactly the same techniques and make the same errors if asked for a more accurate costing. Professionals working in contracting companies do a very detailed design and price all the goods and services required to supply it, consider risks, margins, contingency, and so on (or, if they don't work in a contracting company, they ask a favor of someone who does).

Engineers have (of course!) quantified this into five classes of estimate as given in Table 12.1. These are used by public bodies in the United States and worldwide (see also the AACE Practice Guide in "Further Reading"):

Table 12.1 Classes of cost estimate

Estimate class	Name	Purpose	Project definition level
Class 5	Order of magnitude	Screening or feasibility	0—2%
Class 4	Intermediate	Concept study or feasibility	1—15%
Class 3	Preliminary	Budget, authorization, or control	10—40%
Class 2	Substantive	Control or bid/tender	30—70%
Class 1	Definitive	Check estimate or bid/tender	50—100%

THE BASICS

The most basic point of all is: if you aren't considering price, you aren't doing engineering. Engineers consider the cost, safety, and robustness implications of every choice we make at every stage of a project.

I have worked in a few places where technical and economic evaluation have been split, and all have provided salutary lessons in why they should not be. Decision making processes were very poor, and too easily swayed by fashion (yes, there is such a thing in engineering!) or the whim of managers.

The degree of confidence you have in your technical design is the maximum degree of confidence you should place in your costing. More usually we have to price in all kinds of other risk factors to arrive at a robust pricing. So it isn't just a question of what the kit costs, risk needs to be priced.

You have process risks—the more novel the process, the greater the chance it will underperform, or fail to perform at all. If your plant fails its performance test, your company will probably be paying penalties every day until it is fixed at your company's expense. You can buy performance bonds which insure process risk, but they cost money, and the more novel you have been, the more they are likely to cost.

Then you have financial risks—overseas contracts can be subject to currency fluctuations, and even home contracts can see significant inflation. If you have made heavy use of some material subject to price fluctuation (which need be no more exotic than stainless steel), things can cost a lot more than you expected.

There are political risks—countries can fall out with each other, industries can be nationalized without compensation, wars break out and, closer to home, regulation can disallow certain approaches, or make them—for example waste disposal—far more expensive than you originally costed for.

Sensitivity analysis is the key to understanding these risks, and deciding how to price them. You are unlikely to win a competitive tender if you price all of them in to your offer at 100% probability. A guide to the kind of price which is reasonable would be probability of occurrence multiplied by cost of occurrence. Many of your competitors in a commercial situation will, however, undercut this value considerably.

In commercial practice you need to consider all of these factors, and produce a price accurate to a few percent.

This price will need to be based upon a design which is optimized to meet the client tender evaluation criteria: lowest price that meets the specification, lowest whole-life cost, best net present value (NPV), fastest payback period—all affect every aspect of competitive design.

In one way or another, all design is competitive. Even if you are doing an in-house design, it needs to be the best design it can be against the evaluation criteria, and you can rest assured that when it goes out to the engineering contractor, they will be redesigning it as much as they are allowed to maximize their profit, and minimize their risks.

ACADEMIC COSTING PRACTICE

In order to decide if it is economic to proceed with a design we need a quick way to estimate capital and running costs. The main plant items (MPIs)/factorial method is almost always used in academia (though far less commonly in practice).

Douglas (see later section) offers a more sophisticated process which increases in resolution as the project progresses, namely economic potential. This is, however, frequently misused nowadays—the more rigorous stages he proposes are left out, with the result that costing is reduced to the vestigial stub of a comparison of feedstock and product costs.

Capital cost estimation by MPI/factorial method

We cannot obtain supplier quotations for all of our equipment and engineering services in academia as a professional would, so we need a standalone costing methodology for use in the academic setting.

Chemical Engineering departments worldwide seem to do more or less the same thing.

First, we estimate the cost of MPIs, usually from cost curves: Timmerhaus and Peters (see "Further Reading") contains many of these curves. We then add factors to account for things like operating pressure, special materials and so on, to the base costs for the curves.

Having added up all the MPI costs, we calculate installation and other engineering and construction costs as a percentage of MPI costs, using Lang factors such as those to be found in Chapter 6 of Sinnot and Towler (see "Further Reading"). This allows us to estimate the capital cost (capex) of the plant.

An academic criticism can be made of this near-universal academic approach, enshrined in Sinnot and Towler, the bible of academic design practice. Lang factors date back to the 1930s, and other more accurate factors have since been devised to replace them.

Operating cost estimation

In academia, operating costs (opex) are usually estimated as a percentage of capital costs, often a nominal 10%. It is actually possible to get a lot closer to professional practice than this (even in a university setting) but very few in academia try, as far as I know.

Professionals estimate how much power, chemicals, manpower, capital, and so on will be required to run the plant, and price these inputs at market rates. There is no reason why even the greenest student cannot try this approach, though obviously they will not be as good as an experienced professional. My experience in asking to students to do this is that they are not as terrible at it is as you might think they would be.

Economic potential

Economic potential (EP), as it is explained in Douglas's Conceptual Design of Chemical Processes, does start with a simple comparison of feedstock and product prices, but rapidly advances to a far more sophisticated accounting of costs and benefits than the standard MPI/factorial approach.

The approach makes a number of assumptions which mean that it is only applicable to a subset of process plant types, but is in its appropriate setting superior in my opinion to the MPI/factorial method.

The very sketchiest version of EP (intended only for use before any design has been undertaken at all) is frequently nowadays the sole costing consideration in academic "plant design" exercises.

Payback period, NPV, and so on

A slightly more sophisticated financial analysis can be undertaken in an academic setting, as well as in professional practice.

Payback period tells us how long it takes to get back our capex from revenues/profits. NPV discounts future revenues and expenditure to reflect the fact that we care less about our money in the future than we do about our money now, and also inflation/interest rates on money.

NPV can incidentally be criticized, as large expenditures far in the future are automatically thought fairly unimportant. This can be used to justify projects with very high future decommissioning costs (such as, e.g., oil rigs and nuclear power plants) in ways which green groups disagree with. Accountancy is not value free.

Sensitivity analysis

Even though academic costing methodology is necessarily a bit flaky, we can firm things up (or at least quantify our flakiness) with an honest sensitivity analysis.

Sensitivity analysis varies the costs and revenues which might apply to a system and considers the shape of the curves obtained. If profitability falls off sharply around your assumed costs and revenues, your process economics are not very robust.

I am personally not so bothered about whether students or other beginners get a realistic price, as whether they know how good their price is—give me a range in which the professional price lies, and a realistic estimate of where it is most likely to lie.

PROFESSIONAL COSTING PRACTICE

I spent most of the first five years of my career producing proposals for turnkey plants for design and build contractors in the (ultra-competitive) international water industry. I was pretty good at it by the end, and I used to win quite a lot of contracts for the plants I designed and bid.

This was sometimes based on price and sometimes on technical merit. It isn't always about getting the lowest price on the table—it does usually help a lot, though.

I have been keeping my hand in in the intervening years and little seems to have changed other than that we now make a lot more use of computers and external design consultants that we used to.

Accurate capital cost estimation

Usually, competitive bids are invited from potential suppliers for the various goods and services used to construct a plant before a process contractor makes a firm offer to an ultimate client. Three is usually thought a good number of bids to have for any item. A smaller number means that there might be a limited number of places where that item can be obtained, which is risky.

Bids are checked against the specification, to ensure that all which has been asked for has been included (frequently not the case), and that the requested payment terms and other contract conditions have been complied with (also frequently not the case). Once bids have been standardized, prices are compared, and a supplier is selected on an "or approved equal" basis.

These prices constitute firm offers by third parties to supply the item for a given sum. They are not at this point estimates, they are guarantees to offer the goods for the price quoted.

Enquiry documents need to be detailed enough to allow suppliers to understand completely what is required both technically and commercially. If they are not, suppliers may decline to quote, or may price in the uncertainty.

Purchasing companies will have their own terms, ultimate client companies will have theirs, and equipment vendors will have their own. It is frequently the case that enquiry documents will ask for quotations based on a combination of client and contractor terms, and vendors will offer their own terms in their offers.

This is not a trivial matter, and the differences in prices between alternative suppliers can be less than the price implications of variation in contract terms. This issue will need resolving to obtain a firm price basis.

If you work in a process contracting organization, you may well have access to many such firm prices for exactly the kind of equipment you are pricing from previous jobs. The basis of your estimates can be very accurate indeed.

Bought-in mechanical items

Professional engineers price unit operations as one or more purchased items of equipment (known as "bought in items," i.e., physical plant bought as discrete items) by sending enquiry documents to relevant equipment suppliers.

These prices usually have to have sums added to address the bits the various suppliers have left out of their bids, so that they can be evaluated on a like for like basis.

They will probably also have sums added to reflect risk. For example, the fewer potential suppliers you have, the greater the risk that prices will rise, or that your bit of kit will not be available in time or at all.

Bought-in electrical items

Control panels, aka Motor Control Centers (MCCs) can be bought as a discrete item or along with electrical installation and/or software supply.

It will usually require input from an electrical engineer, and probably an element of in-house design to be able to produce sufficiently detailed enquiry documents to obtain reasonably accurate quotations for MCCs.

PCs, PLCs, DCS systems, or supervisory computers may also be bought as discrete items or integrated with the MCC.

If anything, greater care needs to be taken to adjust bids, and to price risks associated with these bids, than it does with those for mechanical equipment.

Mechanical installation

Mechanical installers will usually supply (in addition to the skilled labor required to fix and mechanically commission the mechanical bought in items) the pipework, bracketry, supports, and so on required to make a working plant. They may also do detailed design of pipework support systems, supply any nonspecialized valves, and so on.

These bids are at best only as good as the drawings the bidders have been issued with, though they are less prone to underestimation and price escalation than electrical installation bids.

Electrical installation

Supply and installation of cables, emergency motor stop buttons, site lighting and small power, and making connections from MCC to motors will normally be the responsibility of a specialist contractor.

This element is possibly the most prone to underestimation by beginners. It is important to issue sufficient information to installers to make sure that everything needed has been accounted for and, ideally, the offer should be checked by an in-house electrical engineer.

Software and instrumentation

This may be provided in-house by some combination of MCC supplier or installation contractor. A specialist may be used to install and commission instruments, program PLCs, and set up SCADA, DCS, remote telemetry, and such systems.

Whoever is doing it, great care has to be taken in pricing this element, as it is a major source of cost overruns at construction stage, especially due to underestimation of the number of inputs and outputs to the system.

Civil and building works

How much is a ton of pumps?

Anonymous Civil Engineer

Civil engineering companies work on very tight margins, and tend to interpret their communications very literally. They work from drawings, so you need to make sure that anything issued to them for pricing is very clearly marked with respect to those elements which you are willing to stand by later, and those which are indicative only.

Their pricing methodology is based on counting tons of stuff. Once they have completed a design, they "take off" from their drawings how many tons of concrete, steel, and so on are required. They are consequently usually in a hurry to get their longest lead time item (design) started, and will pressure you for the required information.

It is best to wait until you have a reasonable degree of certainty before issuing it, if for no other reason than because civil engineering companies have a reputation for being rather more litigious than other disciplines.

Civil and building costs are relatively easy to control as long as you have nailed down the usual weasel words in civil engineering pricing documents ("unforeseen ground conditions" for starters) during the initial stages.

Design consultants

Nowadays, companies are increasingly using the services of design houses to carry out design, particularly for specialist items.

If you are going to do this, you will need to price it in, and allow for the strong possibility of requirements for additional design work later in the project.

This can come to a surprising amount of money. At the time of writing, the going rate in the United Kingdom for an experienced process design engineer is £150 per hour or so.

Project programming

Professional engineers produce a schedule or program of events setting out the time-scales for the key elements of the design/construction/commissioning phase and allocating resources against each of the tasks required.

This allows pricing of those items whose costs are based entirely on their duration of use (such as, e.g., hire of site cabins) as well as indicating how many hours will be required for each discipline, and whether the company has the resource to handle the project in-house, or will need to buy in (usually more expensive) external resources.

Man-hours estimation

The plant design engineer will have produced their estimate of how many hours of each discipline it will require to do the job, but the discipline heads within a company will also want to give their estimate of how long it will take their people to do it.

Since they are the ones who have to deliver the project, and the plant designer is responsible for winning the work, discipline head estimates tend to be on the high side, and plant designers on the low side. There should be some negotiation.

Pricing risk

Once you have prices for all the goods and services you need to make the plant, you need to make sure that you have allowed money toward the chance that process, financial, legal, political, or other risks go against you.

As well as adding sums to individual prices as previously described, you might do this formally by buying a form of insurance known as a performance bond, which usually costs a fraction of a percent of the complete contract value. You might add an overall contingency, which is built into your price. Alternatively you might declare the risk to the client, and include a prime cost (PC) sum which you would charge if the possible adverse event materializes.

Margins

Margins vary greatly from industry to industry. Back when I was pricing water treatment plants for a living in a very competitive sector, we were happy to get paid 22% more than our bought in costs.

Some very sharp practitioners were bidding contracts at less than cost, by leaving things out which had to be included later (under what are called variation orders, VOs) at top dollar.

Generally, the less money there is swilling around in a sector, the tighter the margins will be, and the more sharp the practitioners.

Competitive design and pricing

It's the only kind I know, and this book is based throughout on the assumption that process plant designers are doing it for profit rather than fun (though it is fun when you get the hang of it).

You can cut your margins of safety as far as you dare, you can negotiate with suppliers, discipline heads, and financial directors at the pre-tender stage, but you can only get so far by reducing your bought in cost and margins by either arm-twisting or charm.

The way to win better contracts more of the time is to design yourself ahead. Don't do what everyone else is doing, but a little less well, for a little less money—do something better. That's why process engineers get the big bucks.

You don't need to be too radical to find all sorts of little ways to be a little bit cleverer than the other guy, and if you find enough of them you can win work with decent margins.

Much of it is to do with seeing the system working together as a whole and seeing the full implications of making small changes. It's all about system level design, the subject of the next part of the book.

Accurate operating cost estimation

With a well-developed design, the contractor knows how many operator man-hours are required, and has an idea of what each discipline costs an employer. They know estimated chemical use, and can calculate expected effluent costs with the Mogden formula. They have accurate estimates of hours run for motors, and can forecast the price of electricity. They have maintenance schedules and costed spares lists obtained from suppliers, and so on. The contractor should consequently be able to cost the expected running cost for the plant to a high degree of accuracy.

FURTHER READING

AACE International, 2005. Recommended Practice no. 18r-97: cost estimate classification system—as applied in engineering, procurement, and construction for the process industries 2005 AACE International. <http://www.aacei.org/non/rps/17r-97.pdf>.

http://www.reliabilityindex.com/manufacturer.

Peters, M., Timmerhouse, K., 2002. Plant Design and Economics for Chemical Engineers. McGraw-Hill, New York, NY.

Sinnot, R.K., Towler, G., 2005. Chemical Engineering Design, Vol. 6. Butterworth-Heinemann, London.

PART 4

High Level Design

The previous section was a description of how to design the components and subsystems of a process plant, but the ability to do this is not why process plant designers get the big bucks. We are paid to produce a design integrated at a higher level than this, so that all the subsections work together well.

I have identified three areas where a whole-plant understanding is most important—process control, layout, and safety. This understanding has to be fed back to the subsystem design covered in the last section, but these three areas have to be addressed primarily at the whole-system level.

CHAPTER 13

How to Design a Process Control System

INTRODUCTION

As Myke King (see "Further Reading") has pointed out, much of what is taught in Chemical Engineering courses under the heading of process control is out of date, irrelevant, and impractical, with the result that most new process plant designers have little idea of how to design the process control aspects of their plant.

What a plant designer needs to be able to do is to specify control loops based on instruments and control actions which make the plant approximate steady state under all conditions thought reasonably likely (or, to put it another way, within the design envelope). In order to do this, standard control approaches for unit operations are an excellent starting point.

Myke also advocates having sufficient consideration of process control issues to build controllability into the design, an approach developed more fully by Luyben (see "Further Reading"), albeit in quite a narrow field.

I see the rationale for this but I am not sure that Luyben's formal and simplified approach is the answer. Like so many elements of process design, academic approaches laboriously solve problems which can be better solved by simpler intuitive means.

I can see the need for integrating process design and control, but I would go further, including under this heading things which might not be thought of as process control elements, such as hardware selection, hydraulic design for passive flow equalization, integrated consideration of process control and safety elements, and the interaction both of operating and maintenance manuals/operators, and functional design specifications/software. I give an example from my own experience of how this works in practice at Appendix 1.

Once they start as practitioners, feedback to novice designers from commissioning and control engineers or their attendance at HAZOPs will hopefully eliminate features which lead to poor controllability, but I would like to give newbies more of a head start than is presently usual.

Integration of process control and design by professionals is far more intuitive and qualitative than mathematical. To quote Myke King:

> It's a difficult subject. I've learnt how to design control strategies instinctively. I've been asked on many occasions to document a methodology. I've got as far as "Work in the industry for 40 years and you get the hang of it."

Difficult it may be, but beginners definitely seem, in my experience, to need to be given a place to start. Interactions with more experienced engineers will refine their understanding, but what they need in the first instance is a way, as an absolute minimum, to put the basics on their Piping and Instrumentation Diagrams (P&IDs). That is what this chapter aims to provide.

MATCHING DESIGN RIGOR WITH STAGE OF DESIGN

At conceptual design stage, very little or no consideration needs to be given to process control issues, unless the plant has some novel or very hazardous components which are likely to present entirely new or very high-risk process control problems.

At the detailed stage of design, a fully thought out and instrumented P&ID needs to be produced, and ideally, precise models of instruments specified. As a minimum, realistic instrument choices and specifications should be produced. Instruments can be expensive, and vary between manufacturers in their requirements and capabilities.

Perhaps more importantly, the number of inputs and outputs to the control system cannot be determined unless the instrumentation has been thought through to this degree.

Modern instrumentation tends to be smart, with considerable onboard processing power. We need to decide how we are going to use this. Are we going to have smart instrumentation with dumb control, or dumbed down instrumentation with smart control?

And then there is the question of whether we are going to have a smart plant with dumb operators or a dumb plant with smart operators? Plants tend to be smart nowadays, but I have been asked to design a fully manual plant to be run by postdoctoral researchers.

The software for a dumb plant is defined in the Operation and Maintenance manual, and for smart plants in the functional design specification (FDS). As most plants have some smarts nowadays, a combination of the two documents will be required to understand how the designer thinks the plant will be controlled.

OPERATION AND MAINTENANCE MANUALS

Operation and Maintenance (O + M) manuals are written for (almost) every process plant, describing how it is to be operated and maintained, and how to troubleshoot any problems which occur. They certainly should be written (and they should also be read!).

They are to me largely a type of process control software since, on a fully manual plant, they describe in detail the control actions which people will undertake to achieve the things which a programmable logic controller (PLC) would do on a fully automatic plant.

Most plants are, however, not fully automatic. There are automatic control actions, and there are manual interventions. Some of these manual interventions are required by law. For example, the UK Institution of Electrical Engineers Regulations requires under certain circumstances (such as motor overheating) stopping the operation of a motor in such a way as to require manual intervention to restart.

Some of these conditions are thought trivial enough to allow the system to automatically restart itself via remote command. Some are thought dangerous enough that the system forces someone physically to press a button before restart is possible and the O + M manual tells them that they must to go and look at the kit before they press the button.

So decisions have to be made about safe operation of the plant; and how software and operating procedures will work together to ensure safety.

Control philosophies always, to my mind, make implicit assumptions about how the plant will be operated, which it is better to include in the document, as another reader may make different assumptions if they are not made explicit.

SPECIFICATION OF OPERATORS

The level of education and training of operators and their availability has to be specified to determine the degree of automation which a plant requires.

In choosing whether to have a highly automated plant, one needs to consider the advantage of operators over instrumentation—operators can detect not just specified conditions, but unspecified and unexpected conditions.

The more we expect our operators to do about the things they monitor, the higher their required level of skill and understanding needs to be.

A fully manual plant will need a high availability of highly experienced staff. A fully automatic plant may need no permanent staff on-site at all, especially now that we can access plant telemetry and system control and data acquisition (SCADA) systems remotely via IP technology.

AUTOMATIC CONTROL

We don't build fully manual plants in the developed world nowadays. Computers are too cheap and reliable, and operators too expensive (and human!), for routine operation activities to be best done by people alone.

Control is mostly done using a combination of PLCs, PCs, and high level control system (distributed control system (DCS) or SCADA), though there may be a few field-mounted controllers specified for a number of reasons.

We don't make a lot of use of physical field-mounted PID controllers of the kind still talked about in university process control modules, and the things most like them which we do occasionally use have their own built-in control algorithms.

Process plant designers do not write the software for these controllers. We might to some extent if we were commissioning or control engineers tuning the control loops, but we would probably do the majority of that by plugging in a laptop and pressing the "optimize" button on the manufacturer's dedicated software.

Commissioning (while very important) is not the subject of this book. Process plant designers need to know how to specify instrumentation and control hardware, populate their P&IDs with these items, and write FDSs so that software engineers can design and price their software.

Process plant designers need to have an idea of what neighboring disciplines do, and what they need to do their jobs. We don't, however, need to be able to do their jobs; we have to be broad-brush people. We don't sweat the details.

Specification of instrumentation

Instrument engineers/technicians (aka tiffies) have their specialism, but we don't need to be one of them to specify instruments well enough to design a process plant. Table 10.8 should help you with this.

We should be willing to be corrected by a tiffy at more detailed design stages on details of instrument choice, as we should by other specialists.

It is, however, unlikely that if we are experienced designers, our choice would not have worked at all. The specialist's choice might, however, work a little better (as long as they fully understand what we want the instrument to do).

Precision

Precision in mathematics is (confusingly to engineering students) to do with what engineers call resolution. When I tell my students off for "spurious precision," this is the sense in which I am using it.

For example, in the filter pretreatment example in Appendix 1, I say that we need control of pH to within 0.1 pH units. This is not the same as saying that we need control to within 0.10 pH units, which implies 10 times the (mathematician's) precision.

Precision in engineering is different, and is to do with repeatability and reproducibility. It is not to do with how close the measured value is to the true value (accuracy) or the smallest change in the measured value with the instrument can detect (resolution). It is to do with whether the instrument will give me the same reading against the same true value the next time I test it.

We might further split engineering precision into reproducibility and repeatability, the first encompassing variability over time, and the second being precision under tightly controlled conditions over a short time period.

(I received a few comments from my correspondents about these definitions, and offers of alternatives, so it is fair to say that there is a problem with precision in the language of precision. These are the definitions I choose to use for the purpose of this book. To quote Humpty Dumpty: "*When I use a word, it means just what I choose it to mean—neither more nor less*").

In our specific example, pH probes require regular recalibration against standards to maintain precision and accuracy. Over the period between calibrations, the accuracy (measured value for a given true value) varies. Eventually it is not possible to calibrate the instrument to give accurate readings against the standards, and a new probe has to be substituted. There are gradual decreases in accuracy, precision, and response time during the periods between new probe installations.

Accuracy

Accuracy is to do with the gap between the true value and the value indicated by the instrument. In the filter feed treatment example mentioned above, I need pH to be controlled within the range around the set point $\pm$ 0.05 pH units, therefore my measurement accuracy needs to be reliably at least this good.

Cost and robustness

Instrument precision and accuracy both tend to cost money. Very precise and accurate (basically lab grade) instrumentation also tends to be less robust as well.

Lab instruments tend not to be suited to field mounting. We therefore tend not to specify any more accuracy or precision than we strictly need, and we may take manufacturers' lab test values for an instrument with a pinch of salt.

All instrumentation needs to have a purpose to justify its cost. It may be true that "you can't manage what you don't measure" but it's best not to measure things you don't need to manage.

Safety

Safety-critical instrumentation requires a higher standard of evaluation than that which only affects operability, or less important still, process monitoring without associated control actions.

We might, for example, specify for a process or for a safety-critical reading the use of redundant cross-validated instruments (in which the reading most likely to be correct is determined by a voting system).

Specification of control systems

The P&ID shows graphically, whilst the FDS describes in words, what the process designer would like the software to do. Turning these deliverables into code is the job of the software engineer.

PLCs are the basis of many modern process plant control systems, with DCS or SCADA supporting the control interface or HMI. The petrochemical industry uses PLCs only for low-level distributed control functions, and prefers to use a combination of DCS and a supervisory computer for overall plant control. This allows for their more sophisticated control functions, which are at least as often retrofitted by control engineers as designed into the original system.

The true nature of the control system should be reflected on P&IDs and in FDSs. We should not expect to see a local control loop and field-mounted controller on a P&ID representing a loop which actually works via signals going out and back via PLC. There are appropriate symbols in the British Standard to show this correctly.

STANDARD CONTROL AND INSTRUMENTATION STRATEGIES

In this section I will break down process control systems into some commonly used blocks, which should allow you to populate your P&ID and control philosophy with the standard features which appear on almost every plant.

I will assume that you know what feedback, feedforward, and cascade control are, but that the rest of your university module on process control was taken up with mathematical software engineering stuff about transforms and algorithms. In twenty-first century process control, signal processing is built into the box, and algorithm writing is done by the software engineer, though they may well need input from the process engineer to do with outcomes of control functions.

Commissioning and control engineers who straddle the divide between process engineers and software engineers need a deeper understanding, but their jobs are very little to do with process plant design. Process plant designers do, however, need to understand what software and control engineers are going to need from them, so that they can design in controllability.

Alarms, inhibits, stops, and emergency stops

Process plant designers will need the assistance of electrical engineers to ensure compliance with the IEE regulations and the various European directives which apply to this area.

However, I have included this section because beginners usually do not understand that all electrical equipment needs to be easy to switch off in an emergency, and very frequently comes with safety features that switch it off automatically in a number of potentially hazardous situations. These might include such things as motor winding over-temperature, motor over-torque, fluid ingress, and so on.

It is frequently the case (and it may be a legal requirement) that the more hazardous of these cases will be set by the electrical/software engineers such that they require an operator to attend site to reset the "trip."

Less potentially serious conditions may stop motor operation only while the state is current or, if less serious still, may only prevent the motor from starting. Both of these conditions might be called inhibition.

All of these conditions will usually be set to generate local alarms in software. More serious ones may generate off-site alarms, or activate an alarm beacon on site.

A design which has an excessive number of alarms should be avoided. If there are too many alarms, operators will be subject to alarm flooding, and develop what is known in health care as alarm fatigue and either ignore them or find ways to disable them. So, how many is too many? The Engineering Equipment & Materials Users' Association (EEMUA) suggests the following criteria (See Table 13.1):

Table 13.1 EEMUA criteria for acceptability of alarm rate in steady state operation

Long-term average alarm rate in steady operation	Acceptability
More than one per minute	Very likely to be unacceptable
One per two minutes	Likely to be over-demanding
One per five minutes	Manageable
Less than one per 10 min	Very likely to be acceptable

EEMUA Publication No. 191.

It should be noted that commissioning engineers frequently disable alarms and interlocks during the early stages of commissioning, but this should be a planned aspect of a commissioning procedure, and suitable substitute safety plans should be made.

Many of these signals, alarms, and interlocks will have to be handled by the control system, and leaving them out of the control system specification, if that is the case, is a classic beginner's mistake leading to cost overruns down the line.

Many are, however, wired directly into the motor starter, which eliminates a potential weak link in the chain. Hard-wiring is standard for safety critical interlocks.

European standards also require the provision of emergency motor stop buttons immediately adjacent to motors. Resetting the emergency stop locally cannot incidentally allow the drive to restart, but there has to be a trip to reset on the motor control center (MCC) as well.

Chemical dosing

There can be many nuances to design of dosing pump systems dealing with liquids which release gases on suction, leak detection, overpressure, cavitation, and so on, but I will deal here with the most common issues.

Pump speed control

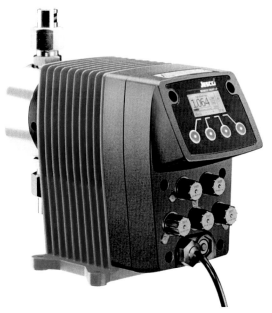

Figure 13.1 Memdos Smart LP: stepper motor controlled dosing pumps offering smoother almost continuous dosing. *Copyright image reproduced courtesy of Alldos.*

There are now digital dosing pumps like the one illustrated in Figure 13.1 with integrated speed and stroke control on board, working from digital inputs originating in a flowmeter and pH probe.

However, it is still common to control a piston diaphragm pump's motor speed with a 4–20 mA signal from a flowmeter in the stream to be dosed. This is known as flow pacing, and when used in conjunction with stroke length control as described in the next section, it can give very accurate (± 0.1 pH units) pH control.

However, some engineers use simple speed control proportional to the difference between measured pH and set point. This is okay, but not as precise as flow-paced stroke control unless the flow into which we are dosing is always constant.

A more old-fashioned way to do this simple type of control is to send pulses to the pump at a frequency corresponding to the desired stroke frequency. Commonly available pumps and pH controllers can usually handle both pulsed or 4−20 mA control signals.

Pump stroke length control

Figure 13.2 Memdos E ATE: mechanically actuated diaphragm dosing pump with inverter controlled motor and actuator/servomotor for automatic stroke length adjustment. *Copyright image reproduced courtesy of Alldos.*

A 4−20 mA signal from a dedicated pH controller can be input to a suitable flow-paced dosing pump to control stroke length, giving a robust two-variable control of chemical dose (Figure 13.2).

Actuated valve control

There are still some plants being built in which an actuated valve is used to add chemicals by gravity into a mixed tank but, to put it very politely, this is a bit old hat nowadays. Control loop time is long, chemical flow control is pretty rough, and the homogeneity at the point of pH measurement is questionable.

Compressors/blowers/fans

Positive displacement

Positive displacement blowers need similar control systems to positive displacement pumps (see later), though the compressible nature of gases makes these systems a little more forgiving than their liquid equivalents.

Centrifugal

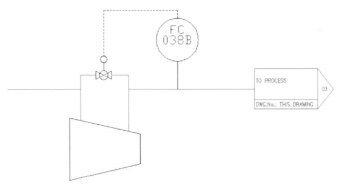

Figure 13.3 CAD representation of centrifugal compressor control.

Centrifugal compressors (Figure 13.3) are more efficient at large sizes than positive displacement blowers, but they are more difficult to control. They are therefore quite often favored where there are high fixed flows.

They are, however, capable of variable output. Back when I started as an engineer, we used to do this on single stage compressors with variable position inlet guide vanes, and on multistage compressors with inlet throttling valves, but nowadays inverter control is usually favored.

Surge conditions—in which too low a flow causes a sudden powerful reversal of flow—have to be avoided, and this is usually achieved via a control valve in a bypass back to the compressor suction. There may be one of these valves for each compression stage, going back to the inlet of that stage.

Such valves may also be used to control flow through the compressor if inverters are thought too expensive. As inverters get relatively cheaper all the time, I would predict that this will eventually become an obsolete approach.

Distillation

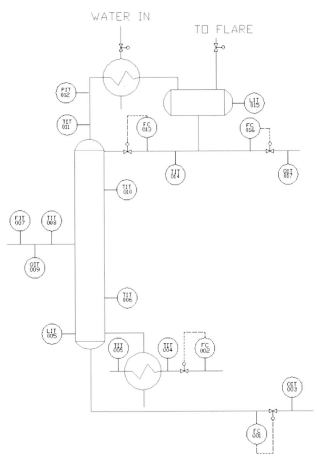

Figure 13.4 CAD representation of distillation column control.

There are many processes and measured variables which we might consider—which of the 120 possible permutations will we pair to which? There is a lot to this, and for more detail I recommend Myke King's book, a chapter of which is dedicated to what he considers a broad outline of the subject. I have illustrated in Figure 13.4 his basic suggestion, which is essentially pressure controlled.

Filters

Backwash control

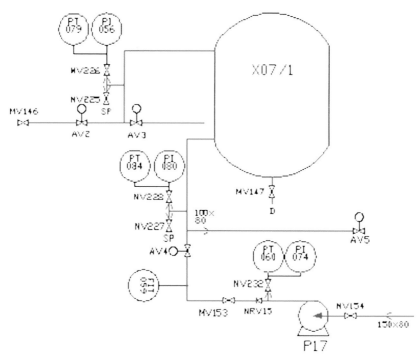

Figure 13.5 CAD representation of backwash control.

The standard methodology for differential head control of particulate filters is that accumulated dirt is removed by reversing flow through the unit, known as backwashing (Figure 13.5). This is done periodically on the basis of a number of criteria:

- Differential pressure (almost always)
- Time since last backwash (almost always)
- Queueing/hierarchy of wash initiation (very frequently if there are filters in parallel)
- Outgoing turbidity (fairly rarely)
- Outgoing particle size analysis (very rarely)

Each filter will therefore need measurement of incoming and outgoing pressure, via separate instruments or a differential pressure instrument. There will usually also be a timer in PLC software (or less frequently nowadays in MCC hardware). Queuing, if required, will be handled in PLC software.

Solids removal efficiency can be measured online via turbidity or particle size analysis. Turbidity measurement is reasonably cheap and robust, though it adds complexity which is usually redundant. Particle size analyzers are very expensive and fragile kit, and best avoided if at all possible.

There will need to be a backwash pump, whose output flowrate is crucial to effective backwashing. An associated flowmeter is therefore required. The backwash flow needs to be high enough to effect dirt removal, though not so high that the filter is damaged, but the acceptable range of flows is usually fairly broad.

It is therefore usually the case that that the commissioning engineer sets this flow by throttling a manual valve or setting a range of inverter outputs, and thereafter it is just monitored and returned to commissioning values by maintenance staff if required. We may, however, sometimes specify a more sophisticated system with temperature-dependent flow control of the backwash pump to ensure a constant mass rather than volumetric flowrate during backwashing.

Chemical cleaning control

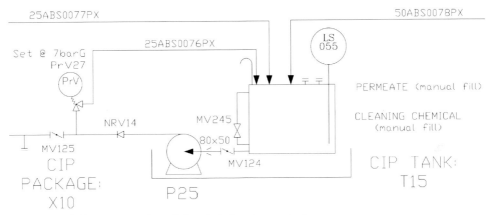

Figure 13.6 CAD representation of chemical cleaning control.

Membrane filters often require a chemically enhanced backwash (CEB) in addition to simple backwashing (Figure 13.6). While the control of this is quite sophisticated, based on analysis of trends in differential pressure across the membranes compared with original condition and the condition after the last backwash, the instrumentation and available control actions are basically the same as for simple backwashing. The modifications to backwash frequency, cleaning chemical type and strength and so on, which are instituted in response to declining membrane performance, are usually manually initiated.

The CEB system's tanks, dosing and centrifugal pumps, flow control, and so on are controlled as described in their respective section of this chapter. If a heated backwash is used, there is a control loop which modulates the output of a process heater in response to a temperature measurement. This loop may well be critical—such systems (most notably the very expensive membranes themselves) are often made of polymers which can be damaged by even quite moderate excessive temperatures.

Fired heaters/boilers

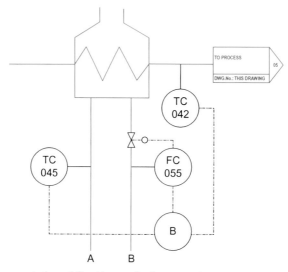

Figure 13.7 CAD representation of fired heater/boiler control.

Fired process heaters have to account for variation in composition of feed, and variation in pressure in the case of gaseous fuels. It may also be the case that the heater feed flow cannot be controlled, as its feed is the product of another process and cannot be economically stored.

One way around this is the dual firing option shown in Figure 13.7 above. The heater duty is set above that of the greatest expected yield of uncontrolled feed gas, and a second fuel is added as required to top up the heater output to the duty required based on the temperature of the fluid being heated.

This is a very complex area, which Myke King discusses in some detail in his book "Process Control: A Practical Approach" from which the above diagram is taken.

As he explains there, boiler control is essentially the same as fired heater control except that the control is via steam header pressure rather than based on heated fluid temperature. He also draws attention to the difference in control requirements for fixed duty (baseload) boilers and the assist (swing) units which are used to control the steam pressure.

Heat exchangers

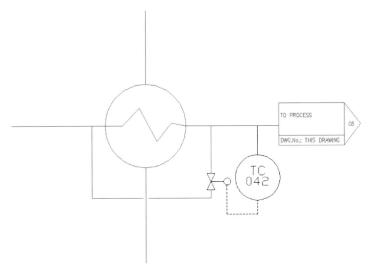

TO PROCESS

05

DWG.No.: THIS DRAWING

TC
042

Figure 13.8 CAD representation of heat exchanger control.

Heat exchangers (Figure 13.8) are usually designed to have temperature sensors in both process and service streams going in and coming out. It is possible to vary the service flowrate to control process stream temperature. Tighter control is, however, given by bypassing the process side with a control valve in the bypass—when this valve is operated, the temperature changes almost immediately.

Pumps

Dry running protection

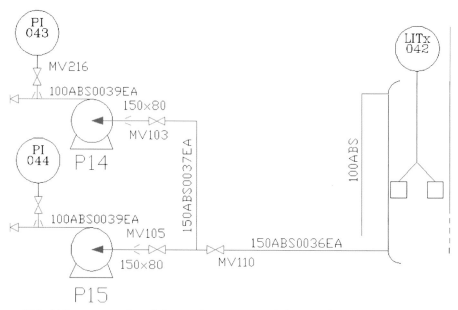

Figure 13.9 CAD representation of dry running protection (of P14/15 by LITx0420).

Many pump types are damaged if they run without liquid for any length of time, so control loops are used to prevent this and are a standard feature of pump configuration (Figure 13.9). The most common variant is an interlock between pump running or starting and the level in a tank feeding the pump, such that low level in the feed tank inhibits pump running and/or starting.

Ultrasonic, radar, hydrostatic, or float type level sensors are most commonly used to provide the level signal. Float switches are very cheap, ultrasonic are good for non-contact measurements of aggressive liquids and powders, and hydrostatic or radar types are good if there is likely to be significant foaming. An interlock can be wired directly into the motor starter from the sensor, or control can go via PLC.

No flow protection

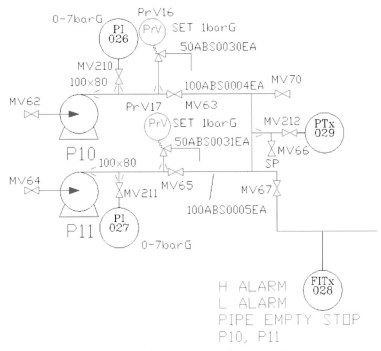

Figure 13.10 CAD representation of no flow protection (by FITX028).

Many modern flowmeters can detect empty pipe conditions, and this can be used as a secondary measure to prevent dry running (Figure 13.10). Alternatively, flow switches can be used for this duty. This is sometimes hardwired into the starter. Going via the control system is, however, probably best so that a timer can be placed in the loop to help commissioning engineers to prevent nuisance trips from transient conditions, especially in the case of flow switches.

Over-temperature protection

Most electric motors come with thermistors incorporated in the windings, so that drives can be stopped automatically if they are getting too hot. This safety critical interlock is usually hardwired into the starter.

Pumps: centrifugal

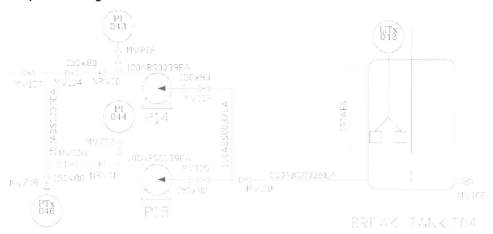

Figure 13.11 CAD representation of centrifugal pump control.

Centrifugal pumps (Figure 13.11) are not immediately damaged by being operated against a closed valve (though they can overheat in fairly short order and even water pumps have been known to suffer steam explosion) but throttling their suction can cause immediate cavitation.

Their output can be controlled by running them against a control valve in the delivery line, or by a bypass valve returning output to pump suction, though I personally prefer to use the more efficient inverter control.

Flow delivered by a rotodynamic pump is inversely proportional to system pressure (though Q/H curve shapes differ), but a flowmeter on the delivery side can be used to control the degree of actuated control valve opening or inverter frequency to accurately deliver the desired flowrate against variable delivery pressure.

Centrifugal pumps offer no resistance to reverse flow (and can generate unwanted electricity if run backwards), and protection is sometimes put in to address this, though nonreturn valves are usually thought sufficient protection if their (approximately 10%) backflow prevention failure rate is acceptable.

Pumps: positive displacement

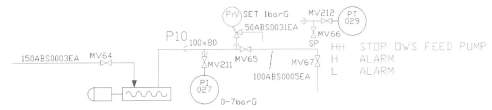

Figure 13.12 CAD representation of positive displacement pump control.

Positive displacement pumps (Figure 13.12) are quickly damaged by being run against a closed valve on the delivery side, and are not usefully controllable by throttling on either side. You can control them to some extent with a valve on a bypass to suction, but very accurate control can be given using an inverter drive.

I would still recommend the use of a flowmeter to modulate the bypass valve position or inverter frequency. Though delivered flow is largely independent on backpressure, wearing parts in these pumps may cause delivery volumes to drop during the intervals between servicing.

These pumps are normally protected from damage caused by valve closing or other line blockage with a pressure relief valve, placed between the pump and the first valve downstream.

Pumps: dosing

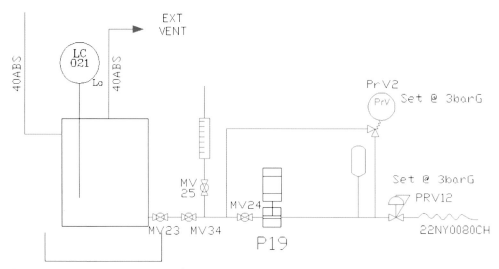

Figure 13.13 CAD representation of a positive dosing pump control.

The most modern digital dosing pumps have on-board integrated stroke speed controls, driven directly by digital signals. Figure 13.13 shows the pressure relief valve (PrV2) which protects against pump damage in the event of line blockage, as well as the pressure sustaining valve and pulsation damper which remove flow pulsations.

Piston diaphragm pumps commonly come with two 4−20 mA inputs for control. One controls motor speed, the other stroke length.

Solenoid pumps are far simpler (and cheaper); their solenoid produces a stroke for every pulse of power sent to it.

There is more detail on this in the "Chemical dosing" section of this chapter.

Tanks

High High High HHH Alarm, Stop P12 and P13
High High HH Start Assist pump
High H Start Duty Pump
Low L Stop Assist Pump
Low Low LL Stop Duty Pump
LOE (X2) Alarm, Stop P12, P13, (P18 AND P19,
P20, P22

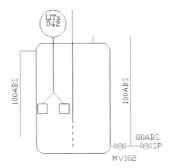

BREAK TANK:T04

Figure 13.14 CAD representation of break tank filling and emptying control.

Though not favored by those who like elegant solutions, and ideally to be made as small as possible, buffer tanks (Figure 13.14) make for a flexible and robust plant design.

Breaking the plant into sections ending/starting with a buffer or break tank is a solution favored by most commissioning engineers, as it makes it easy to commission the plant in sections.

Pumped flows can be ramped up and down, based on levels in the feed and/or delivery tanks, in such a way as to maintain either a fairly constant tank level, or a fairly constant flow.

In either case, rapid changes in flowrate and on/off control of pumps should be avoided. Smoother operation is normally better operation. Level sensors such as ultrasonic and hydrostatic types, which measure level continuously, rather than trip at a threshold level are therefore preferred.

Ultrasonic level indicator controllers can handle this control function well using their on-board electronics.

Valves

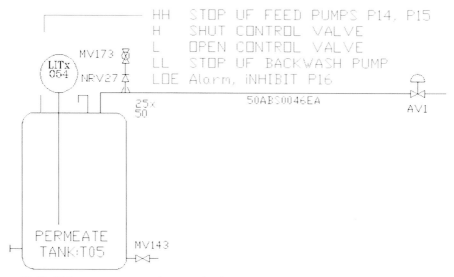

Figure 13.15 CAD representation of actuated valve control.

Figure 13.15 shows how an on/off actuated valve can be used to control the level in a tank. There are some details of the nature of control valves which are not commonly explained in university courses and which I will cover in the sections which follow.

Rotary actuators—modulating duties

Figure 13.16 Rotary actuator. *Copyright image reproduced courtesy of AUMA.*

Globe and other valves used in a modulating duty require multiple controlled turns to go from open to closed, and consequently require multiturn rotary actuators, which are most frequently electrically driven (Figure 13.16).

Butterfly valves go from open to closed in 90° of shaft rotation, and there are "quarter-turn" actuators used to operate them. Such actuators may be electrically or pneumatically driven.

Linear actuators—open/closed duties

Figure 13.17 Linear actuators (painted red). *Copyright image reproduced courtesy of Ascendant.*

There are linear actuators which are used in a vertical orientation with globe and other rising spindle-type valves, usually in on/off applications (Figure 13.17).

Valve positioner/limit switch

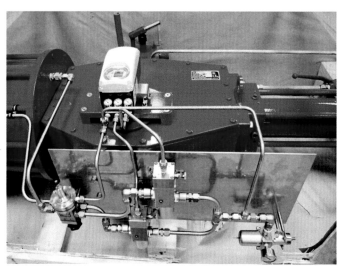

Figure 13.18 Valve positioner/limit switch. *Copyright image reproduced courtesy of Ascendant.*

Valve positioners and limit switches tell the system when the valve has reached a certain position, so that it can be reliably driven to a certain degree of opening (Figure 13.18). This gives a positive indication that the valve has in fact reached the desired position. This is not a guarantee of a given valve headloss or throughput, but it does increase the accuracy of operation considerably.

FURTHER READING

Ali, R., 2013. Keep it Down. The Chemical Engineer, November 2013. Available at <https://www.tcetoday.com/ ~ /media/Documents/TCE/Articles/2013/869/869alarms.pdf>.
King, M., 2010. Process Control A Practical Approach. Wiley, Chichester, UK.
Luyben, W.L., 2011. Principles and Case Studies of Simultaneous Design. Wiley, Chichester, UK.

CHAPTER 14

How to Lay Out a Process Plant

INTRODUCTION

It has been estimated that 70% of the cost of a plant is affected by the layout, and its safety and robustness are if anything even more strongly dependent on good layout.

This area is the most notable omission from modern chemical engineering courses, and it is also the reason why I came to write this book. In introducing (perhaps reintroducing) professional engineering design to the University of Nottingham's degree courses, I found that we used to teach an entire degree module on this subject.

J.C. Mecklenburgh, a former engineering practitioner who taught at the University of Nottingham, wrote an IChemE book to accompany the module which, until recently, was still in use at many other universities. My offer to produce an updated version of that (now out of print) book led to the book you are reading being commissioned. So this is arguably the most important chapter of this book.

I have borrowed heavily from Mecklenburgh's approach throughout this chapter. I wish I had space to include more of his content, as his book is hard to get hold of (at the time of writing, Amazon had a few copies priced at £460 each), and is mostly still current professional practice. I am presently producing an updated version of Mecklenburgh's book, which should be available by 2017 at the latest.

There are a number of aspects to the more or less complete omission of plant layout from today's university courses. There has more generally been a loss of drawing and visual/spatial skills from chemical engineering courses. Many universities (to the extent that they use drawings at all) accept Google Sketch or MS Visio sketches which bear no resemblance to engineering drawings. They restrict even these drawings to Block Flow Diagrams ("BFDs") or half-baked approximations of Piping and Instrumentation Diagrams (P&IDs).

Then there is the tendency to reward abstraction and purism in academia. Lecturers are taught on our teaching courses that the "extended abstract" is the manifestation of the highest sophistication of student submission. Professional researchers need little encouragement to think that realism isn't just unnecessary, it isn't very clever.

Not being very clever is the worst thing of all in a university. There is also the usually unvoiced suspicion which many scientists have that artists aren't very clever.

It is often thought axiomatic by academics that there is a hierarchy of different "intelligences" thus:

logical/mathematical
linguistic
social
spatial intelligence

Of course, all of these are really just secondary issues. The main reason academics don't teach layout and drawing is because they can't do these things themselves. They only seem trivial to people who haven't tried them.

Having carried out design exercises with a wide range of audiences, I can confirm that only the most excellent academics (usually professors) produce plant layouts as good as their best students. Plant layout requires flexing those mental muscles which universities usually do not exercise. So how do novice designers make a start? Let's start with some principles.

GENERAL PRINCIPLES

We need to consider layout in sufficient detail from the very start of the design process. Even at the earliest ages of design, attempting to lay plant out will throw up practical difficulties.

Layout is not just a question of making the plant look pretty from the air (or more usefully from public vantage points). The relative positions of items and access routes are crucial for plant operability and maintainability, and there are more detailed site-specific considerations.

The three key elements which have to be balanced in plant layout are cost, safety, and robustness. Thus, wide plant spacings for safety increase cost and may interfere with process robustness. Minor process changes may have major layout or cost consequences. Cost restrictions may compromise safety and good layout.

The layout must enable the process to function well (e.g., gravity flow, multiphase flow, net positive suction head (NPSH)). Equipment locations should not allow a hazard in one area to impinge on others, and all equipment must be safely accessible for operations and maintenance.

As far as is practical, high cost structures should be minimized, high cost connections kept short and all connection routing planned to minimize all connection lengths.

Layout issues at all design stages are always related to the allocation of space between conflicting requirements. In general, the object most important to process function must have first claim on the space. Other objects must fit in the remaining free space, again with the next most important object being allocated first claim on the remaining space. The constraints always conflict, and the art of layout lies in balancing the constraints to achieve an operable, safe, and economic layout.

Layout most obviously affects capital cost, since the land and civil works can account for 70% of the capital cost. Operating costs are also affected, most obviously through the influence of pumping or material transfer cost and heat losses, but more subtly in increased operator workload caused by poor layout.

The layout must minimize the consequences of a process accident and must also ensure safe access is provided for operation and maintenance. Things further apart are less likely to allow domino effects from explosion, fire or toxic hazards, but it is likely that lack of space means that complete mitigation of fire, explosion, and toxic risk will not be achievable by separation alone.

Layout starts by considering the process design issue of how the equipment items function as a unit and how individual items relate to each other.

For example, the individual items in a pumped reflux distillation unit should be close together for effective fluid flow and minimal heat loss. The condensers and drums should be near ground level to reduce the cost of associated structures, but the drums must be elevated to provide NPSH for the pumps. Such relationships may sometimes be identified by a study of the Process Flow Diagram (PFD) or P&ID, but not all will be as obvious as this example. This is where a General Arrangement (GA) drawing, experience, and judgment become vital to find and balance the physical relationships in the layout.

There are many other factors to consider. Equipment needs to be separated in such a way that it can be safely accessed for maintenance, for other safety reasons such as zoning potentially explosive areas, and to avoid unhelpful interactions.

Exposure of staff to process materials needs to be minimized. Access to areas handling corrosive or toxic fluids may need to be restricted. This may require the use of remotely operated valves and instruments located outside the restricted area.

In a real-world scenario we will have a site or sites to fit our plant on to. Different technologies will have a different "footprint." They will have a required range of workable heights and overall area. They may lend themselves better to long thin layouts or more compact plants.

There may be a choice to be made between, for example, permanently installed lifting beams, davits, and so on, more temporary provision, or leaving the eventual owner to make their own arrangements.

There are no right answers to these questions, but professional designers will need to have given these issues sufficient consideration to be able to argue that they have exercised due diligence. In the United Kingdom, the minimum extent of this due diligence is specified in the Construction Design and Management (CDM) regulations.

Layout does not require complex chemical engineering calculations but it does require an intuitive understanding of what makes a plant work, commonly known as professional judgment. If it were more common, we might call it "engineering common sense."

These qualities cannot be taught formally, but must be acquired through practice. It is, however, possible to start the learning process in an academic setting by design practice, though judgment will mostly be developed in professional practice.

Mastery comes only from learning from experienced practitioners and by testing one's own ideas rigorously. The best way to do this is by listening to those who build, operate, and maintain the plants you design for 15 years or so, and adopting their good ideas (no matter how embarrassing the learning process might be during layout reviews).

Generally speaking, the most economical (and easiest to understand/explain to operators) way to lay out a plant is for the process train to proceed on the ground as it proceeds on the P&ID, with feedstock coming in on one side, and product out of the far side of the site or plot. There may, however, be arguments for grouping certain unit operations together in a way which does not correspond with the P&ID order if the site is on multiple plots with any degree of separation.

Land is cheap in some places and expensive in others. Tall plants are generally more favored away from human habitation, but some places don't mind big ugly plants if they come with jobs attached. Such places might also be happy to have quite dangerous processes close to human habitation, as we were before we could afford to be as fussy as we are now.

FACTORS AFFECTING LAYOUT

The main things to consider in laying out a plant or site are the same as they were in Mecklenburgh's day: Layout is, in short, the task of fitting the plant into the minimum practical or available space so that each plant item is positioned so as to balance the following competing demands.

Cost

- The capital and operating costs must be affordable (e.g., placing heavy equipment on good loadbearing ground).
- The plant must be capable of producing product to specification with the practical minimum levels of operation, control, and management.
- Regular maintenance operations should be capable of being performed as quickly and easily as is practical. Units should be capable of being dismantled *in situ* and/or removed for repair.
- The plant must be arranged so as to promote reasonably rapid, safe, and economical construction, taking into consideration staging of construction/length of delivery period.

Health/safety/environment

- Safe and sufficient outgoing access for operators, and incoming access to close to units for fire fighters needs to be designed in.

- Operating the plant must not impose unacceptable risk to the plant, its operators, the plant's surroundings, or the general population.
- Operating the plant must not impose excessive physical or mental demands on the operators. Manual valves and instruments, for example, need to be easily accessible to operators.
- Design out any knock-on effects from fire, explosion, or toxic release.
- Avoid off-site impact of noise, odor, or visual intrusion, mainly by moving such plant away from boundary and communities, and presenting nice offices and landscaping to public view.
- The plant, whether enclosed in buildings or outdoors should not be ugly through uncaring design, but should blend with its surroundings and should appear as the harmonious result of a well-organized, careful design project reflecting credit on the designer and plant operator.

Robustness

The plant and its subcomponents must be so arranged as to operate and make its product(s) as specified.

- Process requirements (e.g., arranging plant to give gravity flow).
- The plant must be designed to operate at the planned availability and should not be subject to unforeseen stoppages through equipment failure or malfunction.
- Consider how the plant might be expanded in future, and allowing space and connections to do so easily.
- Consider access to commissioning resources in layout.

Site selection

The site layout must provide a safe, stable platform for production over a period which may be measured in decades. It is essential to define, early in the design process, if the site is to be used by the designer or others for a single plant or if several plants are to be installed either now or in the future.

If future plants are planned, some assessment of future development is needed. It might be that space needs to be reserved, road networks planned, and major utility distribution expanded to serve the new plants. When a site specification is drawn up, the site layout aims to make the best use of all features of the site and its environs, for example:

- Site topography
- Ground characteristics
- Natural watercourses and drainage
- Climate
- External facilities
- Water, gas, electricity supplies

- Effluent disposal services
- Transport of people and goods

Due care and attention also needs to be given to the effects of the plant on the site surroundings, especially:

- Housing
- Hospitals, schools, leisure centers
- Other plants
- Forests, vegetation
- Wildlife
- Rivers and groundwater
- Air quality

If we have a number of candidate sites which we need to choose between, the following factors should be considered at a minimum:

- Desired layout of the proposed complex
- Cost, shape, size, and contours of land/degree of leveling and filling needed
- Loadbearing and chemical properties of soil
- Drainage: natural drainage, natural water table, and flooding history
- Wind: direction of prevailing winds and aspect, maximum wind velocity history
- Seismic activity
- Legacies of industrial activity like mine workings, chemical dumps, and in-ground services
- Ease of obtaining planning permission
- Interactions with both present and planned future nature of adjacent land and activities

Many of these are environmental factors. We might split them into three categories: natural, man-made, and legislative.

Natural environment

Weather varies greatly from place to place. Singapore has lightning on average 186 days a year. Cherrapunji in India had 2.5 m of rain over two days in 1995. Death Valley has air temperatures of up to 57°C, and Antarctica down to −89°C. A gusting wind speed of 300 mph was once recorded in Oklahoma City. There may also be sandstorms, earthquakes, tsunamis, floods, snow, hail, fog, and so on to consider.

Might our pipes freeze, or might they be softened by foreseeable ambient temperatures? How much rainwater might we need to handle? What earthquake and wind loadings do we need to specify? All need to be considered from the start.

Well-designed plants are site-specific: process plant designs cannot be cut and pasted from one location to another, as a recent TCE article (see "Further Reading") suggests is occurring in South East Asia.

Man-made environment

At the detailed design stage, we will need to consider the possibility of effects on and interaction with surrounding plants, installations, commercial, and residential properties.

In the United Kingdom, this is likely to involve interaction with regulators such as UK Department for Environment Food & Rural Affairs (DEFRA), the Environment Agency (EA), and Planning Authorities.

Regulatory environment

The likely effluents, emissions, and nuisances (gaseous, solid and liquid as well as noise and odor) and any abatement measures need to be considered at the earliest stage.

It should not be assumed that regulatory authorities will allow any release to environment, or sewage undertakers allow any discharge to sewer without discussion with them.

As well as the question of simple permission, there will be the question of emission quality, which will set the size and cost of on-site treatment, or the ongoing bill for third-party treatment. Not all effluents can be economically treated on site, and third-party costs can be very significant, so this needs to be considered at the earliest stages of design.

In the United Kingdom, third-party costs can be predicted using the Mogden formula:

$$C = R + M + Vm + V + Bv + \frac{OtB}{Os} + \frac{StS}{Ss}$$

R is Reception charge

V is Primary treatment charge (also referred to as P)

M is Treatment charge where effluent goes to a sea outfall

Bv is Biological treatment charge (also referred to as $B1$ and Vb)

Vm is Preliminary treatment charge for discharge to outfalls

B is Biological oxidation charge (also referred to as $B2$)

S is Sludge treatment charge

Os is Chemical oxygen demand of settled sewage

Ss is Suspended solids concentration in crude sewage

UK readers can download a Mogden formula calculator free of charge here: http://www.wrap.org.uk/content/mogden-formula-tool-0

It should be borne in mind that the regulations which cover releases to environment always become more stringent over time. Future-proofing might be considered. If you are not going to include additional plant, you need to at least allow space for such plant.

PLANT LAYOUT AND SAFETY

The IChemE book "Process Plant Design and Operation" (see "Further Reading") contains the following suggestions about safety implications of plant layout:

Incompatible systems should be separated from each other; humans from toxic fluids; corrosive chemicals from low grade pipework and equipment; large volumes of flammable fluids from each other and from sources of ignition; utilities should be separated from process units; pumps and other potential sources of liquid leakage should where possible not be located below other equipment to minimize the chance of a pool fire. (This is particularly important with fin-fan coolers where air movement may fan the flames)

The Health Safety Environment (HSE) states that the most important aspects of plant layout affecting safety are for our designs to:

...prevent, limit and/or mitigate escalation of adjacent events (domino); Ensure safety within on-site occupied buildings; Control access of unauthorized personnel and Facilitate access for emergency services.

So the advice is fairly consistent. A well-run Hazard and Operability (HAZOP) should pick up any issues in this area, but designers should not go in to a HAZOP with ill-considered layouts which will require extensive modification. HAZOP is not supposed to involve redesign. Safety studies should be carried out as part of the design exercise, as described in the next chapter.

As well as these guiding principles, there are many detailed considerations, such as siting dangerous materials as far as practical from populations and control rooms, considering plan operability and maintainability, safe access and loading/unloading of deliveries and collections, and so on.

Regulatory authorities including DEFRA, the EA, and Planning Authorities (or such equivalents as may exist in other countries) may want to place restrictions on the design with respect to siting, distance of certain structures from people, height of certain structures, size of inventory of specified substances, releases to environment, and so on. These issues need to be considered as soon as possible in the design process. Certain types of plants, especially those likely to be covered in Europe by Control of Major Accident Hazards (COMAH) legislation will need special attention.

There are quite a number of codes of practice and guidance notes with respect to plant layout as well as many specific codes for things like installations handling liquid chlorine. A good place to start with these, especially for UK readers, is the HSE site (see "Further Reading"). The Dow/Mond indices are also explained there—these can be used at an early design stage as rules of thumb to give outline guidance on equipment spacing. I include more international references in the next chapter, which is specifically on safety.

The general principles designers most need to bear in mind are Inherent Safety and Risk Assessment.

Mecklenburgh gave the following principles for separation of source and target:

- Large concentrations of people both on- and off-site must be separated from hazardous plants.
- Ignition sources must be separated from sources of leakage of flammable materials.
- Firebreaks are wanted, and are often best provided by a grid-iron site layout.
- Tall equipment should not be capable of falling on other plant or buildings.
- Drains should not spread hazards.
- Large storage areas should be separated from process plant.
- Central and emergency services should be in safe areas.

He also produced the tables of suggested separation distances to be found in the appendixes. These are of course for guidance only, and should be subject to the application of engineering judgment.

PLANT LAYOUT AND COST

As far as capital cost is concerned, the further apart things are, the more piping and cable is required to connect them. Things further apart also require more steelwork and concrete, land and buildings to support and contain them. Smaller plant often costs more than less space-intensive plant, but land costs money too.

With respect to operating costs, things further apart have higher fluid transfer headlosses, higher cable power losses, and higher heat losses from hot and cold services. It also takes more time to go from one part of a larger plant to the next.

There will be a balance to be struck between capital and operating costs, as designers need to think of how every item of equipment will be maintained, how motors and other replaceable items will be removed and brought back, and how vehicles required by operational and maintenance activities will access the plant during commissioning and maintenance activities as well as during normal operation.

A few cost saving guidelines are as follows:

- Buildings should be as few in number and as small as practical.
- Gravity flow is preferred, and failing that pump NPSH should be minimized.
- The number and size of pipes and connections should be the minimum practical.
- Save space and structures wherever possible (consider safety!).
- Group equipment where practical and safe.
- Make multiple uses of structures, buildings, and foundations.
- Make full use of the height available.

- Consider column locations when laying out equipment in a building/choosing building type.
- Don't bury services under buildings.
- Storage tanks can be made of welded steel, bolted glass on steel panels, site cast concrete, preformed concrete, plastics, and composite materials. All should be considered, and have implications for layout with respect to construction access.
- Ground conditions can affect economics of concrete tankage—it might most economically be fully in-ground, fully out, or intermediate depending on the water table and how hard it is to dig, etc.
- Ground conditions can also affect civil costs of locating heavy structures on a site. Put them where the ground is good.
- Structure loading can interact with ground conditions—heavy structures can have a low loading by being shorter and wider. This can be in tension with the process design if a tall thin structure is required.

Having considered the most practical issues, we need to remember that we may need to make our plant acceptable to the public (and their proxies, the planning authorities) if it is to be built. It matters what it looks like.

PLANT LAYOUT AND AESTHETICS

Aesthetics can be very important to engineering designers, but it is more or less absent from engineering curricula. Aesthetics falls within the province of philosophy, or maybe psychology, which is bad news for any hope of coming up with any definitive answers. Philosophers have yet to find a definitive answer to any of the questions which they have been pondering for a couple of thousand years now.

Personally, I like the look of process plants, and I think that great big flames coming out of the top of flare stacks look very cool. However, this is a minority opinion (though it is one shared by Ridley Scott, who incorporated the appearance of the Wilton chemical plant near his boyhood home into his film Bladerunner). Most people don't like the appearance of process plants. Neither do they like how they sound and smell, the associated vehicle movements, or the effect on their house prices.

Architects can help us with aesthetics, though this may come at a price: here's a sewage pumping station designed by an architect for the London 2012 Olympics (Figure 14.1).

Figure 14.1 London 2012 Olympic park pumping station. *Copyright image reproduced courtesy of the Olympic Delivery Authority.*

This will usually be another issue settled by negotiation. Generally speaking, we would like our buildings to be unadorned big metal sheds, and planners (and the general public) may ideally like them to be ancient oasthouses saved from demolition by being lovingly rebuilt on our site to contain our process plant.

This seemingly straw-man example actually comes from my past experience—I recall one job where, after negotiation, we ended up with a row of token cowls of the sort seen on oasthouses on a big tin shed, more than a hundred miles from the nearest real oasthouse.

All of this aside, we will need to care about aesthetics, because planning authorities do (and some would say that good engineering is rarely ugly). The cost implications of building something to the standards required where we will attract NIMBY ("Not in My Backyard!") resistance might make it more economical to build it where people care more about jobs.

Architects will know the rules under which planners have to work, and the ones who know how big a brick is are our best source of information on how to satisfy the planners at least cost. The artist/philosopher type of architect will be far less helpful to us than the other kind. It might be an idea to ask any architect you are considering using how big a brick is (215 mmL × 102.5 mmW × 65 mmH is the right answer) and save yourself a lot of heartache.

Practical architects may also help us to produce a plant which is a more pleasant and efficient place of work. Staff morale is important—some even think it improves performance.

MATCHING DESIGN RIGOR WITH STAGE OF DESIGN

Conceptual layout

Conceptual design aims to do enough work on layout to ensure that the proposed process can fit on the site available, and to identify any layout problems which need to be addressed in more detailed designs.

From a regulatory point of view you need to establish whether enhanced regulatory environments like COMAH or Integrated Pollution Prevention and Control (IPPC) are going to apply to the plant, and design your layout to suit.

There are a number of issues which might be neglected at early stage design by beginners.

In which direction is the prevailing wind?

Upwind/downwind positions can matter, for example, pressure vessels should not be downwind of vessels of flammable fluids, toxics should not be upwind of offices/personnel, and cooling towers need to be as far away as possible from anything which would interfere with airflow, and so arranged as not to interfere with each other.

Indoor or outdoor?

You also need to decide on whether you want indoor or outdoor plant. Good lighting, ventilation and air conditioning, protection for excessive noise, glare, dust, odor, and heat help staff to do their work well, so indoor is often the most comfortable for operators. Indoor is more secret, easier to keep clean, and indoor equipment usually has a lower IP rating, making it cheaper to buy. Outdoor plant is, however, usually best for toxic and flammable vapor dispersal, and buildings cost a lot of money.

Construction, commissioning, and maintenance

Designers need to consider all stages of the plant's life, and specifically vehicle access and space for removal and laydown of plant and subcomponents like heat exchanger tubesheets.

There may also be a requirement for tankers and temporary services for maintenance, commissioning, and turnaround activities. This should be identified and accommodated early in design.

Maintenance activities also need to be covered—instruments and plant requiring regular attendance and maintenance should be reachable by short, simple routes from the control room. (This is particularly important with batch processes, which often require a lot of operator intervention.)

Where equipment is to be maintained *in situ*, space needs to be left for people and tools to reach equipment for inspection and repair, the lifting gear required, and laydown for parts, and the design needs to consider accessing different levels in the plant via ladder or staircase, and how to get tools to the working level.

Where it is to be maintained in the workshop, space should be allowed for people and tools to reach equipment for inspection, disconnection and reconnection, the lifting gear required for removal and replacement, and loading/unloading onto transport to workshop or off-site.

In either case, the working area should be designed to be safe with respect to access, lifting, confined space entry, electrical and process isolation, draining, washing, and so on.

Items requiring regular access for operation, maintenance, or inspection should not be in confined spaces or otherwise inaccessible.

Materials storage and transport

The size of on-site materials storage and transport facilities will be determined in the first instance by practical issues of import and export from site of feedstocks, products, and waste material. It should be noted that there may be a choice of road, rail, and water transport to be made. Larger volumes of transport may be best handled by rail or water transport, if practical.

The process's requirement for storage and transport facilities will need to be moderated by statutory and commercial standards and codes of practice, as well as planning authority requirements.

There is commercially available software that allows the designer to overlay a vehicle turning circle over the road layout to check that the roads are suitable for the proposed use, and I offer guidance in the following section on suitable road and turning circle sizes which can be used in the absence of such software.

Materials storage and transport considerations will form part of the Hazard Assessment (HAZASS), and will probably need to be at least justified and possibly reconsidered as part of planning and permitting applications.

The provision of utilities such as steam, compressed air, cooling, and process water and effluent treatment facilities needs to be considered at the earliest stage, as their design needs to be well integrated with that of the whole plant.

Emergency provision

Mecklenburgh gives relevant dimensions for fire appliances to allow designs of suitable roads, hardstanding, etc. to ensure suitability of access provision for emergency use. To summarize, roads need to be at least 4 m wide, suitable for 20-ton vehicles with turning circles of at least 21 m. About 5 m of hardstanding needs to be provided 5–10 m from buildings and items which might require firefighting.

Security

As soon as equipment is on-site, we will need a site fence to protect company and staff property from theft, and prevent unauthorized access for safety reasons. We may want, in some cases, to design out features which will shelter protestors at the site entrance, or make plant buildings amenable to occupation.

Central services

Admin, welfare facilities, labs, workshops, stores, and emergency services need to be well-sited, and should feature on designs from an early stage.

Earthworks

Earthworks may be required to shield tanks of dangerous materials or controls from the site fence. If we are close to communities, we might also want to consider bored teenagers with air rifles. I once had a site where we had to put bulletproof shields on all emergency shutdown switches visible from the fence, to protect them from target practice, and another where we had to erect sight screens during commissioning to stop them shooting at anyone in hi-viz.

Conceptual layout methodology

Professional engineers and companies have their own ways to doing this, but Mecklenburgh's method for initial plot layout is okay (if perhaps a bit OTT) and goes roughly as follows:

- Generate initial design, sizing, and giving desired elevations of major equipment.
- Carry out initial HAZASS (see next chapter) or apply MOND index, consider all relevant codes and standards.
- Produce plan layout of plant based in this data and the suggested spacing in this chapter (I cut out scale shapes from paper and arrange them on a big sheet of graph paper before I go to CAD for this).
- Question elevation assumptions, consider and cost alternative layouts.
- Produce simple plan and elevation GAs of alternatives without structures and floor levels.
- Produce more detailed plan drawing based on decision for last stage.
- Use this drawing to consider operation, maintenance, construction, drainage, safety, etc. Consider and price viable looking alternative options.
- Consider requirement for buildings critically. Minimize where possible.
- Produce more detailed GA in plan and elevation based on deliberations to date.
- Carry out informal design review with civil engineering input based on this drawing.
- Revise design based on this review.
- HAZASS the product of the design review. Determine safe separation distances for fire and toxic hazards, zoning, control room locations, etc. Consider off-site effects of releases.
- Revise design based on HAZASS.

- Confirm all pipe and cable routes. Informal design review with electrical engineer would be helpful.
- Multidisciplinary design review considering ease and safety of operation, maintenance, construction, commissioning, emergency scenarios, environmental impact, and future expansion.
- Reconcile outputs of design review and HAZASS, taking cost into consideration.
- If they will not reconcile, iterate as far back in these steps as required to reach reconciliation.

Detailed layout methodology

The detailed design stage for plants of any significance should include a number of formal safety studies as well as less formal design reviews which should address layout issues as well as the process control issues which may form the core of such studies.

If it is identified that the site will comprise a number of plots, interactions between these plots and with any existing ones on the site and those on surrounding sites need to be considered at detailed design stage.

Mecklenburgh's detailed design procedure bringing together plot designs (as outlined in last section) into a site-wide design for multiplot sites is approximately as follows:

- Compile the materials and utilities flow sheets for piping and conveyors as well as vehicle and pedestrian capacities and movements on- and off-site.
- Lay out whole site, including areas for the various plots, buildings, utilities, etc.
- Use the flow sheets to place plots and processes relative to each other, bearing in mind recommended minimum separation distances, sizes, and areas.
- Add in services where most convenient and safe from disasters.
- Next, place central services to minimize travel distance (but considering safety).
- Now consider detailed design of roads, rail, etc. keeping traffic types segregated, and maintaining emergency access from two directions to all parts of the site.
- Identify and record positional relationships between parts of the plant/site which need to be maintained during design development.
- HAZASS site layout, with special attention to the possibility of knock-on effects.
- Single discipline design review (Chemical; Electrical; Civil; Mechanical) of design and construction functions, which should critically review the design from the point of view of each discipline.
- Multidisciplinary design review. The various disciplines should critically examine the design with respect to: Hazard containment, Safety of employees and public; Emergencies; Transport and Piping systems; Access for Construction and Maintenance; Environmental Impact including air and water pollution; and Future expansion.

- If there is still more than one possible site location at this point, the candidate sites can be considered in the light of the detailed design, and one selected as favorite.

Reviews at this stage may be undertaken in a 3D model in the richer industries such as oil and gas, but many industries are still doing it with 2D hard copy drawings.

For construction

After site purchase, detailed design to optimize the site to its chosen location can be undertaken. Mecklenburgh has a number of optimization steps earlier in the process, but premature optimization is unwise, so I have dropped them.

Mecklenburgh has another stage of design review and optimization for the "for construction" phase, which basically involves gathering very detailed data on the site, the market for the products, and so on and testing the design assumptions from previous stages for a good match to the real site. He then recommends repeating the hazard and design reviews, culminating in consideration of the plant with its wider surroundings.

FURTHER READING

Anon, Undated. Plant layout. Health and safety executive, <http://www.hse.gov.uk/comah/sragtech/techmeasplantlay.htm>.

Eades, J. (2012) It Couldn't Happen Here. The Chemical Engineer 857, November 2012, pp. 26−28.

Mecklenburgh, J.C., 1985. Process Plant Layout. Institution of Chemical Engineers, London.

Scott, D., Crawley, F., 1992. Process Plant Design and Operation: Guidance to Safe Practice. Institution of Chemical Engineers, London.

CHAPTER 15

How to Make Sure Your Design Is Reasonably Safe and Sustainable

INTRODUCTION

Why is Health Safety Environment (HSE) the number one concern? Process engineering (especially water process engineering) saves far more lives than the best medical practitioners ever will, but we can do harm on an industrial scale as well.

Back when I applied to be a chartered engineer, the only aspect of professional practice in which experience had to be demonstrated in all applications submitted was "safety aspects of process plant design and operation." I consider the subsequent removal of this requirement to be a mistake, as you can become a chartered chemical engineer now without having experience in this area. You cannot, however, become a good engineer without this experience.

The worst doctor who ever lived might have killed a few hundred people over a long period of time. Bad engineering could kill tens of thousands of people in a day.

The most pressing argument for the prime importance of safety issues is the more or less universal ethical one that people should not die or be injured so that we can make money. The argument for avoidance of environmental degradation is weaker. Many societies are willing to put up with a degree of environmental degradation in order to industrialize and develop, just as the industrialised West did. The IChemE metrics reflect the engineer's views on this, which are a fair bit more rational than the views of some environmental pressure groups and anticapitalist protesters.

WHY ONLY REASONABLY?

Engineers make decisions on the basis of more or less formal cost benefit analyses. We know that as we try to move toward perfect safety and sustainability, each incremental improvement becomes progressively more expensive.

This fact is accommodated by UK and European law by the terms "as low as reasonably practicable" (ALARP) and the even uglier acronym SFAIRP (so far as is reasonably practicable), which define the required standards of safety. These terms set a limit on how far we have to go. We are not required to make plants any safer if the cost of an incremental increase in safety is grossly disproportionate to the benefit gained.

An Applied Guide to Process and Plant Design

© 2015 Elsevier Inc.
All rights reserved. 217

So society tells us at least how safe they would like our plants to be through legislation. There may, however, be conceivable situations in which our interpretation of our moral obligations as professional engineers requires us to set a higher standard.

In the United Kingdom and Europe in general, there is legislation which requires higher levels of scrutiny of health, safety, and environmental aspects of a design under specified circumstances. For UK designers, the Environment Agency's "netregs" website and the HSE's are very good places to start looking at the requirements in more detail. The most important aspects for process plant designers are as follows:

Control of Major Accident Hazards (COMAH) legislation requires that businesses holding more than threshold quantities of named dangerous substances *"Take all necessary measures to prevent major accidents involving dangerous substances ... Limit the consequences to people and the environment of any major accidents which do occur."* There are tiers within the legislation which impose higher duties on companies holding greater quantities of these materials. Plant designers need to consider whether their proposed plant will be covered by this legislation at the earliest stages.

Control of Substances Hazardous to Health (COSHH) legislation requires risk assessment and control of hazards associated with all chemicals used in a business which have potentially hazardous properties. Consideration of the properties of chemicals used as feedstock, intermediates, and products is a basic part of plant design. Inherently safe design requires us to consider these issues at the earliest stage.

There is a lot of similar legislation worldwide, and I would recommend the US Center for Chemical Process Safety's publications intended to assist process plant designers in addressing safety issues (see "Further Reading").

Plants which are more safe or sustainable than society requires are unlikely to be built in the normal run of things. I have seen a few cases where this has been done for marketing purposes by companies with enough spare cash not to worry too much about the costs, but mostly plants are built to make stuff in a cost-effective, safe, and robust manner.

MATCHING DESIGN RIGOR WITH STAGE OF DESIGN

We should consider inherent safety (designing out risks) at the very earliest stages of design. Chemists are notorious among chemical engineers for devising process chemistry optimized for their batch/bench-scale processes, and for having a slightly gung-ho attitude to safety issues.

Their greater tolerance of hazardous chemicals is understandable, as they work with far lower quantities than process plants contain. Their bench-scale chemistry is, however, rarely optimal for full-scale production.

CONCEPTUAL DESIGN STAGE

It would be unusual to go out of our way to design in sustainability, but we always design in safety from the very start. This tends in earlier design stages to be informal and instinctive, and progresses to more formal studies later in the project.

Formal safety studies

We aren't going to be able to do a formal Hazard and Operability (HAZOP) at the conceptual design stage due to lack of project definition and documentation, but we can and should carry out less formal Hazard Identification (HAZID) and Hazard Analysis (HAZAN) studies as well as other industry-specific types of formal study appropriate for this job.

It is very important to determine as soon as possible whether the plant is going to come under enhanced regulatory regimes such as IPPC and COMAH. These will have a major impact on plant design development construction and operation costs, and may rule out certain locations and approaches entirely.

Inherent safety

Rather than controlling hazards, we should design them out of our process from the very start. Inherent safety is a way of looking at our processes in order to achieve this. There are four main keywords:

Minimize

Reduce stocks of hazardous chemicals (Trevor Kletz called this intensification, which others later confusingly used to mean something else entirely).

Substitute

Replace hazardous chemicals with less hazardous ones.

Moderate

Reduce the energy of the system—lower pressures and temperatures generally make for lower hazards.

Simplify

KISS! (Keep it simple stupid)

Kelly Johnson

Don't design plants you don't understand, and especially don't pile safety features one on top of another instead of solving the root problem.

Note that the principles of inherent safety are applied at conceptual design stage to the proposed process chemistry.

In the situation where we are being given the process chemistry by a product development team, we need to consider whether they have considered the constraints of full-scale operation. Are the selected reactants, solvents or process conditions the most inherently safe ones? If they are not, and we are in a position to influence the process chemistry, can we get the chemists to rethink the chemistry?

In the more common scenario where the technology/process chemistry is bought in from a third party, can we choose another bought-in process which is more inherently safe?

Human factors

Some of the worst accidents ever were caused by what might be called human error, or more correctly by plants which were not designed with real operators and managers in mind.

The IChemE are very keen on this issue, and in their latest discussion document on the future of chemical engineering it is stated that "The crucial role of human factors in process safety will also shape the institution's approach to process safety."

There are lots of interesting books on the subject of how people interact with process plants—Trevor Kletz's are very readable.

I personally learned a lot by commissioning plant, watching, and training operators. In summary:

- Think about the limits of human attention
- Think about the limits of human physical capabilities
- Specify minimum competence of operators required
- Design your plant so that it is easy to do the right thing and hard to do the wrong thing
- Design your plant so that even if the wrong thing is done, disaster does not ensue
- Design your plant so that it is physically impossible to do truly disastrous things
- Design in controlled operation of your plant with a combination of operating procedures and control philosophy
- Consider carefully limits on access to operating software

Ideally we should, as a profession, build a knowledge-base of how difficult plant situations were handled successfully in the past—being good is more than just not being bad! The IChemE used to keep an accident database, but no longer. When I ask my students if they have heard of Flixborough or Seveso, very few have.

Those who cannot remember the past are condemned to repeat it

Santayana

User-friendly design

Trevor Kletz identified a related concept to inherent safety, which addresses human factors he called user-friendly design. There are a number of additional principles and suggestions which are mostly amplifications of the four principles of inherent safety.

Tolerate errors

There will always be errors in plant operation. If such errors readily lead to disaster, your design is neither inherently safe nor robust. Design your plant to handle these events, ideally passively.

Limit effects/avoid knock-on effects

If there is a chance of hazardous events in plant operation, their effects on people and environment must be minimized by designed features. It is especially important to minimize secondary effects caused by damage accrued by the initial event. Note that this applies only to hazards which it has not been possible to design out—it does not contradict the four main keywords. For example, if tanks containing flammable liquids have a weak seam around the roof, the lid may blow off, but the tank will not rupture spilling the contents.

Make incorrect assembly impossible

Because what can happen will happen, and if the effect of incorrect assembly is significant hazard, it needs to be designed out.

Making status clear

Obvious visual clues as to the status of plant help prevent accidents. To borrow Trevor Kletz's example, commissioning engineers use blanking plates to shut off process lines—a variety of such plates known as a spectacle plate make clear from a distance the status of a line.

Ease of control

Controllability is a desirable feature of process plant designs in its own right, though the safety case is perhaps the strongest justification.

DETAILED DESIGN STAGE

There will be formal safety reviews during detailed design, such as HAZOP, electrical equipment zoning, and other industry specific analyses.

Note that formal does not mean quantitative-qualitative judgments by professional engineers will very likely be the best way to pick up all significant risks until design for construction stage. Quantitative methods applied before those late stages are both overkill and a waste of resources.

These formal methods are the subject of the next part of the chapter.

FORMAL METHODS: SAFETY

In increasing order of rigor we have HAZID, HAZAN, and HAZOP. We also have what Mecklenburgh calls "Hazard Assessment," (let's call it "HAZASS") a design review/risk assessment exercise focusing on safety aspects of layout, which falls somewhere between HAZAN and HAZOP.

HAZID

The first step in management of risk and hazards is to identify potential hazards. This is the purpose of HAZID studies.

Personal safety

Broadly there are five classes of personal safety hazards:

Physical/mechanical
Slips, trips, and falls; confined spaces; noise; burns, cuts, and strikes; heat/cold stress/dehydration.

Biological
Release of hazardous organisms.

Chemical
Release of hazardous chemicals.

Electrical
Including static electricity.

Psychosocial
Poor plant design and management can cause physical and mental health problems for workers. Many of these are potentially lethal to individuals, so consideration of them should not be beneath a process plant designer's notice.

HAZASS

This is a kind of design review, so it is best done in collaboration between a number of engineering disciplines. The Mecklenburgh methodology is as follows:
- Define the process design with drawings (Piping and Instrumentation Diagram (P&ID), General Arrangement (GA), Equipment datasheets, etc.).
- Identify sources of failure and vulnerable targets, and try to design out hazard by removing failure mode or, failing that, path to target.

- Identify parts of plant containing dangerous materials which would be hazardous to release even if no failure mode is identified.
- Estimate the frequency of release and potential rates and amount of leakage.
- Evaluate the potential consequences of release on targets.
- Adjust layout and/or design and repeat assessment until consequences of loss are acceptable.
- Make emergency response plans based on the residual risk, and revise layout accordingly.

HAZOP

I will base this section on what I teach in academia, which is quite a bit more realistic than what is usually taught there. The difficult thing about HAZOP is not in any case grasping the methodology; it is being able to imagine the consequences of failure states, work effectively as part of the HAZOP team, and stay on task (and awake!) throughout the procedure.

It would be usual for new engineers to be trained before attending a HAZOP, usually a one-day course for simple attendance at a HAZOP, which would normally offer a fair amount of practice at participating in mock HAZOPs. There are also longer courses for HAZOP chairs, and experienced chairs are often brought in from outside to lead company HAZOPs.

If you would like to explore HAZOP further than the outline I cover here, I would recommend the late Trevor Kletz's books on the subject (see "Further Reading") which are as readable and entertaining as they are informative.

Process plants are complex, and even the most experienced engineer cannot tell at a glance all the ways in which the parts might interact. HAZOP is a formal technique which allows us to consider how the plant we have designed operates under a number of sets of operating conditions.

This is especially telling in an academic setting where steady state conditions may have been overemphasized—there is no steady state in the real world, and to the extent that we approximate it, it is a product of good process control.

Consideration of the construction, commissioning, decommissioning, start-up, and shutdown phases should be incorporated into the HAZOP.

Commissioning may involve the use of chemicals not used during normal operation (e.g., for pipe pickling), the production of noise, dust, odor, and fumes, heavy traffic loads with knock on effects like mud and dust transfer offsite, as well as a requirement for temporary storage, office, welfare, and sanitary facilities.

During start-up and shutdown, commissioning and maintenance activities the systems responsible for safe operation may be absent. Is our plant safe—not just under our expected operating conditions, but under the reasonably foreseeable conditions at all stages of its life?

HAZOP and other similar tools allow us to systematically evaluate this question. In a university we can only approximate the professional practice, but I will flag up those important areas where we have had to lose an important aspect of the real-world procedure in order to teach it in an undergraduate chemical engineering course.

We will take a set of engineering drawings and supporting documentation, and apply a basic HAZOP methodology to identify possible safety and operability issues. The documents we will consider will be a P&ID, GA drawing, functional design specification, and a set of datasheets.

In professional practice, we would have a larger set of documents available to us, and we would have a multidisciplinary HAZOP team. The inclusion of electrical, software, and mechanical engineers as well as operating and commissioning specialists would maximize the chance that all impacts are considered. In this academic exercise, we will inevitably miss some of the issues these other disciplines would have raised, but we are learning a methodology, rather than carrying out an actual HAZOP.

The basic HAZOP method is to inspect the P&ID one node at a time, and to permutate parameters (flow, pressure, composition, temperature, etc.) with guidewords (more, less, no, reverse, other) in order to identify hazard (to health and safety) and operability issues (which might impact on profitability or the environment).

Nodes are usually selected as sections of the P&ID surrounding and including a unit operation, encompassing those process lines which are most likely to be affected by the unit operation, or to affect it.

There are a number of roles in a real HAZOP team which require specialist training, but I simplify this in academia to just two fixed jobs: a Chair to keep things on track, and a Scribe to note down your findings.

Professional HAZOP procedure

1. Core team: Chair, Scribe, Process Engineer, Control Engineer plus other engineering disciplines and specialists in operation, commissioning, etc.
2. Consider as a minimum four scenarios: Start-up, Steady State Operation, Shutdown, and Maintenance.
3. Go through P&ID one node at a time using guide words (NONE; MORE OF; LESS OF; PART OF; MORE THAN; OTHER) for each parameter (flow; pressure; temperature; component/impurity; phase; viscosity, etc.).
4. Make no assumptions. Chair should be asking "how do we know that ...." to break all assumptions.
5. Record all deviations identified.
6. Identify changes to plant or methods which make deviation less likely or protect against the consequences.

7. Decide if cost of changes is justified.
8. Agree justified changes, agree who is responsible for them.
9. Produce action list with responsibilities.
10. Follow up to ensure action has been taken.

Reporting

The conditions leading to these potential problems and the issues themselves are listed, alongside the actions proposed to mitigate them, on an action report similar to that used in a professional context (Figure 15.1).

HAZOP REPORT SHEET

Name	Title	Role	Sign	Project title	
				Project No.	
				ELD No.	
				Sheet	
				Date	
No.	Guideword/ parameter	Possible cause	Consequence	Action	Person responsible	Date to be completed	Completion signature

Figure 15.1 HAZOP action report sheet.

In the academic exercise, the column identifying who is responsible for making sure the corrective action is done is omitted. In the real world, this is very important. Things which are everyone's responsibility are no one's.

In addition to P&IDs, the GA is an important tool to help put the cause/consequence scenarios identified by the team into a plant context and visualize the possible hazard location. When a consequence is identified, the P&ID does not show which other items are in the vicinity and might be affected. The GA allows the team to make a rapid judgment of the potential for knock-on effects.

Similarly, if a Quantitative Hazard Analysis is required, the hazard contours can be drawn on the layout to show where unacceptable hazard levels are imposed on equipment and access areas.

Professional HAZOP is a lengthy and expensive procedure not to be undertaken lightly because of the unavoidable cost of the rigor which gives its results their value. It requires quite a number of senior engineers, with prior training and experience in the technique. Computer-aided design (CAD) tools are, however, becoming available to emulate the procedure, which might offer real assistance to the HAZOP team.

HAZAN

HAZAN is an initial screening technique which has a number of variants. The general principle is that a table is drawn up for the various hazardous events which might conceivably happen on a plant, with likelihood of occurrence on one axis and severity of outcome on the other, and the product of likelihood and severity is called risk. Dow and Mond indexes are commonly used to rank hazards.

As with HAZID, this is done in accordance with the principles of Process Safety: major hazards are considered as a priority.

Risk matrix

A common approach to risk assessment and hazard analysis is the risk matrix. The underlying idea is that acceptability of risk is a product of how likely a thing is to happen, and how bad it would be if it did (Tables 15.1 and 15.2).

These are combined to produce a risk matrix as in Table 15.3.

Table 15.1 Risk matrix: categorization of likelihood

Category	Definition	Range (failures per year)
Certain	Many times in system lifetime	$>10^{-3}$
Probable	Several times in system lifetime	10^{-3} to 10^{-4}
Occasional	Once in system lifetime	10^{-4} to 10^{-5}
Remote	Unlikely in system lifetime	10^{-5} to 10^{-6}
Improbable	Very unlikely to occur	10^{-6} to 10^{-7}
Inconceivable	Cannot believe that it could occur	$<10^{-7}$

Table 15.2 Risk matrix: categorization of severity of consequences

Category	Definition
Catastrophic	Multiple loss of life
Critical	Loss of a single life
Marginal	Major injuries to one or more persons
Negligible	Minor injuries to one or more persons

Table 15.3 Risk matrix

Likelihood	Consequence			
	Catastrophic	**Critical**	**Marginal**	**Negligible**
Certain				
Probable				
Occasional				
Remote				
Improbable				
Inconceivable				

Key:
• Red: Class I—Unacceptable
• Orange: Class II—Undesirable
• Yellow: Class III—Tolerable
• Green: Class IV—Acceptable

Functional safety standards

Functional safety standards nonexhaustively include provisions for:
• SIL determination
• Project management
• Architecture/redundancy
• Probability of failure on demand
• Equipment selection
• Software
• Proof testing
• Verification & validation
• Audits
• Assessments
• Management of change
• Competency

They are performance based rather than prescriptive; it is essentially a matter of demonstrating fitness-for-purpose throughout the life of the installation. Although a variety of certificates are offered (with varying credibility) by various parties for different aspects of compliance there is no requirement within the standards for certification of anything.

A key point to bear in mind is that the "SIL" is nominated for each individual function, that is, the required effects to suppress the hazard associated with a given cause. A high pressure or high temperature may cause the same valve(s) or drives(s) to trip; these would typically constitute two separate functions; one temperature, one pressure. The trips may cause a variety of additional actions but it is only those that are necessary for the suppression of the hazard that constitute the safety "function."

Note also that protection functions are typically required to be implemented independently of the control systems, the failure of which may give rise to a demand on the protection function.

Safety integrity level/LOPA

LOPA is a HAZAN tool very commonly used in certain industries, which works to quantify the risks associated with hazards identified in a HAZOP exercise in a more sophisticated way than the Risk Matrix.

The approach is founded in standards to do with specification of the required functional safety standards for electrical/electronic/programmable electronic safety-related systems.

The functional safety standard IEC 61508 and its process sector specific derivative IEC 61511 detail the approach to be employed throughout the design, implementation, operation, and maintenance of such systems.

Compliance with these standards is not mandatory, but they are held to represent good practice. The performance requirements of the functions are tied to the risk reduction factor (RRF) target identified for them and are allocated to one of four "Safety Integrity Levels" (SILs):

SIL 1 RRF >10 to ≤100
SIL 2 RRF >100 to ≤1,000
SIL 3 RRF >1,000 to ≤10,000
SIL 4 RRF >10,000 to ≤100,000

In practice, SIL 1 and 2 are relatively straightforward to meet and will be substantially achieved through the application of historical good practice for such functions.

SIL 3 is distinctly more challenging and will almost certainly require some redundancy in the system provision, for example, 1 out of 2, 2 out of 3 voting.

SIL 4 is next to impossible to comply with. If a SIL 4 requirement is identified the likelihood is that (a) your process design is seriously flawed; look to enhance inherent safety and (b) your approach for identifying the risk reduction requirements ("SIL determination") is wrong or incorrectly calibrated.

Trips and Interlocks are known as Instrumented Protection Functions (IPFs) which may be used to help reduce the risk associated with process hazards.

A SIL 1 requirement is routine; perhaps 10−20% of trip functions. SIL 3 is quite exceptional; less than 1% of trip functions. Many IPF have a risk reduction requirement of less than a factor 10 and therefore are *not* SIL rated.

SIL determination may be undertaken by a variety of approaches, for example, risk graph, risk matrix, layer of protection analysis, fault tree, which use more or less rigor in the assessment of risk, the risk reduction contribution from other provisions (e.g., relief valves), and the acceptable (tolerable) level of risk associated with the hazard to be protected against.

Although compliance with the SIL target is largely a matter for the instrument discipline, the identification of the SIL target is very much a process concern. There will typically be a range of instrument system design options and effective management of these issues requires a dialogue between the instrument and process disciplines.

This is potentially fraught territory; provisions within the standard are often cited out of context or without due consideration of the particular circumstances. It is all too easy for an "expert" to weave a plausible but wrongheaded argument. Absolute compliance is something that we approach asymptotically; there typically comes a point where the marginal gain in integrity does not warrant the additional expenditure in resources.

FORMAL METHODS: SUSTAINABILITY

Sustainability is a highly politicized word, but we chemical engineers know exactly what sustainability means, because the IChemE has helpfully told us not just what it is, but how to quantify it, in their Sustainability Metrics. As engineers we understand that if the company goes bust, our business plan was not sustainable, so there are metrics in the IChemE document which measure this aspect of sustainability, as well as some of the fluffier ones.

IChemE metrics

To quote the sustainability metrics document (see "Further Reading"):

The metrics are presented in the three groups
3.1. *Environmental indicators*
3.2. *Economic indicators*
3.3. *Social indicators*
    which reflect the three components of sustainable development.
    Not all the metrics we suggest will be applicable to every operating unit. For some units other metrics will be more relevant and respondents should be prepared to devise and report their own tailored metrics. Choosing relevant metrics is a task for the respondent. Nevertheless, to give a balanced view of sustainability performance, there must be key indicators in each of the three areas (environmental, economic, social).
    Most products with which the process industries are concerned will pass through many hands in the chain Resource extraction—transport—manufacture—distribution—sale—utilization—disposal—recycling—final disposal.
    Suppliers, customers and contractors all contribute to this chain, so in reporting the metrics it is important that the respondent makes it clear where the boundaries have been drawn.
    As with all benchmarking exercises, a company will receive most benefit from these data if they are collected for a number of operating units, over a number of years, on a consistent basis. This will give an indication of trends, and the effect of implementing policies.

A note on ratio indicators

Most of the progress metrics are calculated in the form of appropriate ratios. Ratio indicators can be chosen to provide a measure of impact independent of the scale of operation, or to weigh cost against benefit, and in some cases they can allow comparison between different operations. For example, in the environmental area, the unit of environmental impact per unit of product or service value is a good measure of eco-efficiency. The preferred unit of product or service value is the value added..., and this is the scaling factor generally used in this report. However, the value added can sometimes be difficult to estimate accurately, so surrogate measures such as net sales, profit, or even mass of product may be used. Alternatively, a measure of value might be the worth of the service provided, such as the value of personal mobility, the value of improved hygiene, health or comfort. But a well-founded and consistent method of estimating these 'values' must be presented

The metrics are calculated under the following headings

Resource usage—Energy; Material (excluding fuel and water); Water; Land

Emissions, effluents and waste—Atmospheric impacts; Aquatic impacts

Economic indicators—Profit, value and tax; Investments

Social Indicators—Workplace; Society

So the engineer's approach is (as ever) one of quantifying as best we can and then balancing costs and benefits. We do not set the value of all environmental goods to infinity, and the value of a company staying in business to zero.

SPECIFICATION OF EQUIPMENT WITH SAFETY IMPLICATIONS IN MIND

Introduction

There is an excellent and concise treatment of the principles of safety in design in "Process Plant Design and Operation," which I do not propose to replicate in full here, but there is a useful introductory statement and a few overarching principles:

The design should ensure a secure containment system. It must be robust and capable of handling both over and under-pressure condition plus temperature excursions where appropriate.

The design should avoid one event setting off a larger event. . ..If the process handles flammable materials the sources of ignition must be kept to a minimum. It should be tolerant of small fires and designed to minimize the frequency of large fires and/or explosions In the case of corrosive fluids the design should be tolerant of corrosion both inside and outside the containment.

Principles

Personal and process safety

Much public discussion of health and safety issues (and many daytime TV adverts) focus on personnel/personal safety issues like slips, trip, and falls and the like. In many cases there is legislation which guides us as to how to design out these personal hazards, though many (including IChemE President and HSE Chair, Judith Hackitt) are now arguing that this kind of health and safety legislation is being commonly misused by ambulance chasing lawyers and lazy public officials in a way which is bringing it into disrepute.

"Process Safety," however, tends to focus on the small subset of these risks which have the potential for very serious incidents in industries handling large quantities of hazardous materials.

Release of large quantities of toxic substances, major fires, and explosions are very serious issues with the potential for multiple fatalities. These are the ones which are usually the primary focus of process plant design safety exercises.

Access

A common fault of beginner's designs is a lack of provision of safe permanent access to equipment. This is normally done via platforms and walkways made of open mesh decking, and vertical and inclined ("ship's") ladders, all usually made of galvanized mild steel.

These items are also available in glass reinforced plastic for chemical resistance, as well as aluminum for expensive shininess (but on the one occasion I had a client who specified aluminum instead of galvanized mild steel, it was all stolen the night after delivery).

General
- Manways should be 0.5 m in diameter minimum, and placed facing gangways. Provision should be made for winching a man out.
- Doors should be 0.6 m wide minimum.
- There should always be two escape routes for operators, especially at the top of tall columns.

Horizontal access
- Platforms should come with toeboards and 1 m high handrails. Platforms, walkways and stairways should not be obstructed by pipes or equipment up to a height of 2.25 m.
- Design loads on decking are specified in BS45492 as:

Light Duty (1 person)	$3.0 \, \text{kN/m}^2$
General (regular two-way pedestrian traffic)	$5.0 \, \text{kN/m}^2$
Heavy duty (high-density pedestrian traffic)	$7.5 \, \text{kN/m}^2$

Vertical access
- Minimum height between floors should generally be at least 3 m, and minimum headroom under piperacks, cable tray, and so on not less than 2.25 m.
- Intermediate steps are required for elevation changes over 400 mm.

- Stairs are preferred over ladders for main vertical access, and ladders should be hooped over 1.5 m, and ship's ladders should usually be avoided. (Note that this order of preference is, as is so often the case, in descending cost order.)
- Maximum ladder height without a landing: 7.5 m.
- Ladders should be arranged so that users face into equipment, not out into space, and they should not be attached to the supports for hot pipes, to avoid distortion by expansion forces.
- A clear 1 m square should be allowed on the plan layout for a ladder.

Flammable, toxic, and asphyxiant atmospheres
Explosive atmospheres: DSEAR
The Dangerous Substances and Explosive Atmospheres Regulations (DSEAR) require risk assessment and, ideally, elimination of hazards associated with flammable and explosive substances. The most important aspect of this legislation for the plant designer is to do with classification of areas where explosive atmospheres may occur. This has a major impact on both equipment specification and plant layout.

DSEAR and other directives and standards require that equipment and chambers which may feasibly contain explosive atmospheres as a result of gases, vapors, mists, or dusts be "zoned" based on the probability of occurrence of an explosive atmosphere.

The probability is usually assessed qualitatively, but for those who really like numbers, HSE gives approximate figures for zoning gas/vapor/mist hazards as follows:
- Zone 0: Explosive atmosphere for more than 1000 h/year
- Zone 1: Explosive atmosphere for more than 10, but less than 1000 h/year
- Zone 2: Explosive atmosphere for less than 10 h/year, but still sufficiently likely as to require controls over ignition sources.
  (The corresponding dust classifications are Zones 20, 21, and 22, respectively.)

Ignition sources have to be controlled within these zones. This may require the exclusion of certain types of equipment, or the use of special "ATEX-rated" drives and so on. (ATEX ratings code 1, 2, and 3 correspond to Zones 0, 1, and 2, respectively.)

If there is residual risk of explosion, consideration needs to be given to provision of blast walls, safe paths for discharge of relief vents, explosion-hardened plant and control buildings, and design of tanks and other equipment to withstand explosion.

We tend, when designing plant, to need numbers to work with in order to quantify risks. In the case of flammable and toxic hazards, we have the upper and lower flammable/explosive limits for a material (and its flash point) and a range of workplace exposure limits for acute and chronic exposure to toxic substances defined as follows:

Flammability hazards
- Lower Explosive Limit, LEL/LFL—The minimum concentration of vapor in air below which the propagation of flame will not occur in the presence of an

ignition source. Also referred to as the lower flammable limit or the lower explosion limit.

- Upper Explosive Limit, UEL/UFL—The maximum concentration of vapor in air above which the propagation of flame will not occur in the presence of an ignition source. Also referred to as the upper flammable limit or the upper explosion limit.
- Flashpoint: The minimum temperature at which a liquid, under specific test conditions, gives off sufficient flammable vapor to ignite momentarily on the application of an ignition source.
- Flammable liquids are classed based on flashpoint as:
  - Extremely flammable—Liquids which have a flashpoint lower than 0°C and a boiling point (or, in the case of a boiling range, the initial boiling point) lower than or equal to 35°C.
  - Highly flammable—Liquids which have a flashpoint below 21°C but which are not extremely flammable.
  - Flammable—Liquids which have a flashpoint equal to or greater than 21°C and less than or equal to 55°C and which support combustion when tested in the prescribed manner at 55°C.
  - Inflammable—confusingly for nonnative speakers, inflammable means the same thing as flammable (or perhaps even extremely flammable). Its use should therefore be avoided.
- The higher up this list a substance is, the more we should seek to substitute it with something less flammable (or failing that the more precautions we would have to take).

Toxic hazards

The long-term exposure limit (LTEL) is the time-weighted average concentration of a substance over an 8-h period thought not to be injurious to health.

The short-term exposure limit (STEL) is the time-weighted average concentration of a substance over a 15 min period thought not to be injurious to health.

The HSE Publication EH40 gives exposure limits for a wide range of chemicals (see "Further Reading").

If we identify excessive exposure to toxic chemicals in our design, we should first consider substituting the materials which produce toxic hazards. Failing that, we can use engineering controls such as ventilation, avoidance of enclosure, controlling access to contaminated areas, and so on.

Note that there are many substances which are both toxic and flammable, and both hazards should be considered simultaneously.

There may be some substantially enclosed areas which may have flammable, toxic or asphyxiating atmospheres which we cannot design out. These are classified as confined spaces, and access to them has to be tightly controlled.

Confined space entry

According to the HSE, "*A confined space is a place which is substantially enclosed (though not always entirely), and where serious injury can occur from hazardous substances or conditions within the space or nearby (e.g. lack of oxygen).*"

It is most often the case that the relevant hazard is the possibility of presence of asphyxiating, flammable, or toxic gases, or the absence of oxygen, so there is some overlap with explosive area zoning.

Entering such spaces (even quite shallow trenches can qualify, as an operator can bend down and place his head in the hazardous atmosphere) has the potential for multiple fatalities, and formal risk assessment is required by law before any entry. This will often require any operators entering such a space to have special training and equipment. Entering confined spaces (even if supposedly only for a moment) is a big deal. It is a bad idea to have any equipment requiring operator access in a confined space.

If this cannot be designed out, the safest kind of confined space is one with a direct drop straight from the surface to the working area. Having to navigate turns and level changes in a confined space is a very risky operation—few except mine rescue teams have the necessary skills and equipment.

Lockable covers on confined spaces are a good idea, and it might be best not to have an internal access ladder. Here is an example of an undesirable layout I came across (Figure 15.2):

Figure 15.2 Example of poor layout, including hanging cables and nonrecommended access ladder.

In this example, properly trained staff could have been winched in, and the absence of a ladder would dissuade untrained staff from just popping down to look at something in a way which has led to many deaths in the past. Multiple fatalities have occurred on many occasions in which untrained operators have gone to the rescue of others who went before them, and are overcome by the same conditions. In Qatar, in 2012, seven expatriate operators were killed in a single incident of this nature on the first day I worked there, and there have since been multiple fatality incidents in the United Kingdom.

Wet/dusty atmospheres: ingress protection (IP) ratings

Equipment needs to be specified so that it is suitable for its environment with respect to particle and water ingress. The most commonly used standard is the IPXX standard, where the first X represents a solid particle ingress standard, and the second X a water tightness standard. 0 is no protection, and 5 is dust protection in the case of the solids standard, 8 is immersion below 1 m in the case of the water standard.

Submersible equipment needs to be rated at IP68 or better, and control panels at IP55 or more. Indoor equipment may be rated as low as IP22 (the standard for domestic power sockets), protected only from fingers and water drips over a short period.

SPECIFICATION OF SAFETY DEVICES

No safety device is 100% reliable, so the use of safety devices is only indicated where it has not been possible to design out hazards, which is always the preferable option.

TYPES OF SAFETY DEVICE

Overpressure protection

Inexperienced engineers have a tendency to do one of two things, either they put pressure relief valves (PRVs) everywhere or (much worse) do not include them where they are needed.

The first error adds significant cost to both capex and opex and possibly produces a net decrease in safety through increased complexity. The second error can be a disaster waiting to happen.

I do not intend to go into the detail of relief valve sizing, about which a whole book can be written; instead I will cover some of the basic scenarios in which a PRV could be required.

The real skill of sizing relief valves is not in grinding through the standard sizing calculations, it is the application of engineering judgment to identify the scenarios in which over pressurization could occur, and determining the reasonable worst-case relief load in such an event.

This section is written with reference to API520 and API521, the Oil & Gas Industry standard for PRV sizing (see "Further Reading"). In the United Kingdom/Europe, the Pressure Equipment Directive will need to be complied with in all industries, but the API standards contain some useful rules of thumb for PRV application which have no equivalent in UK or European standards. API521 discusses some common scenarios in which an overpressure (and therefore a "relief case") may occur. The most common of these are as follows:

Closed outlets (on vessels)

In the event that all the outlets from a vessel are closed off (perhaps due to manual valves being closed through an operator error or automatic valves failing), system overpressure may occur. While this is quite unlikely, as engineers we have to assume that all the outlets to a particular vessel might be closed if it is physically possible.

The key issue is to determine if the highest achievable pressure in the vessel is above the design pressure. In many cases this may mean comparing the maximum-rated pressure of upstream pumps or supply vessels to the vessel in question.

Inherently safe design implies that design pressures throughout the entire system which might be over pressurized by such an event are consistent. For example, the rated pressure of the upstream pumps will be equal to the pressure of the downstream vessels/valve, etc.

Burst tube case

Heat exchangers will almost always contain fluid at higher pressure on one side of the tubes than the other, so a burst tube will result in the high pressure fluid leaking into the low pressure side (including, in some cases, flashing of the fluid) which might ultimately cause a catastrophic failure.

In this instance it must be noted that we are not protecting against cross-contamination, but protecting the exchanger/pipework against catastrophic failure and consequent loss of containment.

We can "dismiss this safety/relief case" (i.e., create an inherently safe design) if the test pressure on the low pressure side is higher than the design pressure on the high pressure side.

Cooling water/medium failure

Cooling can be used deliberately to create a pressure drop within a system. In these cases a loss of the cooling medium may lead to increased pressure (a similar scenario can occur through loss of reflux cooling).

An inherently safe design will design vessels for the maximum pressure achievable in the event cooling is lost.

Blocked in (hydraulic expansion)

This scenario often occurs with liquids in heat exchangers in two situations:

1. A cold liquid side of a heat exchanger becomes blocked in while the hot side continues to flow. Depending on the temperature difference, increased heating may cause expansion or vaporization leading to overpressure.
2. Liquid in a line may be blocked in (e.g., by an operator closing a valve in error). If the liquid is normally below ambient temperature (or it has trace heating) it may expand on heating and cause overpressure. While the expansion may be small, in the case of incompressible fluids the pressure can quickly increase and cause a problem.

Exterior fire case

In the event that a fire occurs immediately outside a vessel, the contents will be heated and can over-pressurize the vessel. This is a very difficult case to dismiss, however it can be dismissed if the vessel is at least 7.6 m above the base of the fire, if the vessel is protected by fire-resistant insulation, or in some other way.

Pressure relief valves

PRVs (Figure 15.3) are spring loaded valves which open automatically at a set pressure, releasing the contents of a pipe or vessel to atmosphere or to a vessel depending on design detail. While in theory they should not, in practice PRVs tend to leak increasingly over time so they are not the best choice where complete containment is crucial.

Figure 15.3 Pressure relief valve.

Bursting discs

Bursting discs (Figure 15.4) are an engineered metal plate fitted across a pipe which bursts at a specified pressure, allowing the pipe contents to pass to atmosphere or vessel. They are a better choice than PRVs where containment is crucial, but once burst, they need to be replaced. They are quite often specified as protection upstream of a PRV to prevent fugitive emissions.

Figure 15.4 Bursting disc. *Copyright image of the Safe-Gard Bursting Disc reproduced courtesy of Elfab.*

Blowout panels, etc.

Typical gas/air or dust/air explosion overpressures are of the order of 10 bar. It may not be practical to design vessels to withstand the overpressure (Figure 15.5).

Figure 15.5 Blowout panel. *Copyright image reproduced courtesy of Elfab.*

Instead, in a similar manner to bursting discs, the roof or wall of a building or vessel can be engineered to fail first, diverting a blast in a safe direction, and minimizing damage within the protected space.

It is common practice for the roof of fixed roof atmospheric storage tanks to have deliberately weak seams for this purpose.

Under-pressure protection
Vacuum relief valve
The lids of large tanks such as those used for storage of products and intermediates on oil and gas facilities may only be designed to withstand pressure, and may be readily imploded by surprisingly small degrees of vacuum. They are therefore usually protected by vacuum relief valves (or combined "vent/vac" or "relief/vac" valves) (Figure 15.6).

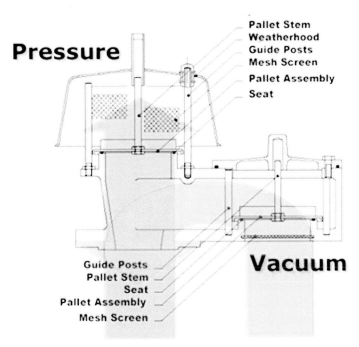

Figure 15.6 Vacuum relief valve.

Static protection

Nonconducting fluids such as hydrocarbons flowing rapidly through pipes or strongly agitated in vessels can produce sufficient static electricity to self-ignite by spark discharge. Contrary to popular belief, metal pipes are actually more likely to exhibit such charging behaviors than nonconductive materials (Figure 15.7). Though the risk may be reduced by reducing liquid velocities, such an approach is unlikely to be reliable or cost-effective enough for complete elimination of risk. Such systems need to be safely earthed to prevent fires caused in this way.

© Newson Gale Ltd.

Figure 15.7 Static protection measure. *Copyright image reproduced courtesy of Newson Gale.*

Gas detectors

Where toxic or flammable gases may be present, permanent gas detection and alarm systems may be required to ensure personnel and plant safety (Figure 15.8).

Figure 15.8 Gas detector heads. *Copyright image reproduced courtesy of Crowcon.*

Emergency shutdown valves

Where it is desired to cut off flow of a component quickly in a potentially hazardous circumstance, valves may be installed which reliably shut off flow in that condition (Figure 15.9). These are known as shutdown valves (SDVs) or emergency shutdown valves (ESVs). They are common features in the oil and gas industry and other safety critical industries.

Figure 15.9 Emergency shutdown valve. *Copyright image reproduced courtesy of Ascendant.*

They are actuated valves, which introduces a number of risks to reliability. The hazardous condition has to be detected, a signal has to pass to the valve, and the valve actuator has to work. All of these things may need to happen very reliably in a condition where the plant is on fire, and main plant power is offline.

For this reason, ESV actuators are normally of the spring return type or actuated by fail-safe fluid power systems, and any signals wiring is fireproofed.

Flare stacks

When PRVs are lifted by overpressure on a gas processing facility, it is undesirable to vent large quantities of flammable gas to atmosphere. Burning the gas in a flare stack makes it safe (Figure 15.10).

Figure 15.10 Flare stack at the Shell Haven Refinery, UK.

Flare stacks are also used to handle gas produced during maintenance and repair activities, plant bypasses, and so on, as well as gas which is considered economically nonviable to recover.

Scrubbers

An alternative way of removing dangerous (usually toxic, nonflammable) substances from a vented stream is through the use of emergency scrubbers (Figure 15.11).

Figure 15.11 Scrubber.

Water sprays

Fixed water spray cooling systems are commonly provided on the tanks used to store flammable hydrocarbons in petrochemical facilities (Figure 15.12). There is a commonly used standard, "NFPA 15: Standard for Water Spray Fixed Systems for Fire Protection," but the design of such safety critical systems is best left to specialists.

Figure 15.12 Water spray system in operation. *Copyright image reproduced courtesy of Lechler.*

Quench tanks

We can arrange for the contents of a vessel containing a reaction which might run away to be dumped quickly to a tank whose physical or chemical conditions stop the reaction very quickly.

FURTHER READING

Anon, 2008. Guidelines for Hazard Evaluation Procedures Center for Chemical Process Safety.

Anon, 2008. Inherently Safer Chemical Processes: A Life Cycle Approach Center for Chemical Process Safety.

Anon, 2011. EH40/2005 Workplace Exposure Limits Containing the List of Workplace Exposure Limits for Use with the Control of Substances Hazardous to Health Regulations 2002 (as amended). Health and Safety Executive.

Anon, 2012. Guidelines for Engineering Design for Process Safety Center for Chemical Process Safety.

API STD 520-1, 2014. Sizing, Selection, and Installation of Pressure-Relieving Devices in Refineries: Part I—Sizing and Selection (9th ed.). American Petroleum Institute.

API STD 521, 2014. Pressure-Relieving and Depressuring Systems (6th ed.). American Petroleum Institute.

Azapagic, A., 2002. Sustainable Development Progress Metrics Recommended for Use in the Process Industries. Institution of Chemical Engineers, London.

Scott, D., Crawley, F., 1992. Process Plant Design and Operation: Guidance to Safe Practice. Institution of Chemical Engineers, London.

SOURCES

Figure 15.6: Free image reproduced from Solids Wiki, http://solidswiki.com/index.php?title=Vacuum_Relief_Valves.

Figure 15.10: Image reproduced under Creative Commons License (http://creativecommons.org/licenses/by-sa/3.0/). Taken from http://en.wikipedia.org/wiki/Gas_flare#mediaviewer/File:Shell_haven_flare.jpg.

Figure 15.11: Image reproduced under Creative Commons License (http://creativecommons.org/licenses/by-sa/3.0/). Taken from http://upload.wikimedia.org/wikipedia/commons/8/87/G_G_Allen_Steam_Plant%2C_scrubber.JPG.

PART 5

Advanced Design

Design optimization proceeds iteratively by stages in professional practice, with an increasing degree of multidisciplinarity, checking, attention to detail, and quality control. This process is quite distinct from academic process integration or optimization techniques.

CHAPTER 16

Professional Practice

INTRODUCTION

Engineering is a collaborative human activity. Humans vary in their technical, intellectual, social, and verbal capabilities, though engineers may not encompass the full range of variation.

While there may be a few individuals with high levels of ability across the range, good collaboration allows those with high technical ability but poor social or verbal abilities (you know, nerds) to complement those with less technical ability, but more charisma and communication skills (the managers of the future). The most creative designers may be rather questioning of authority and intolerant of rules, for the same reasons as they are good at finding creative solutions and are consequently often freelancers.

There are a number of formal interactions between engineers which facilitate communication with those who may be uncomfortable with unstructured conversations. Those which count as work might broadly be termed design reviews, negotiations, and formal procedures, though there are crossovers between these categories.

There is another type of interaction which is very useful to engineers, follows a well understood format, but doesn't usually count as work: discussing what not to do based on personal anecdotes.

Harvey Dearden has produced a little book which helpfully sets out quite a lot of the unwritten rules of professional engineering culture, and which I would recommend to all new engineers (see "Further Reading")—though you can skip the Jane Austen chapter.

GENERAL DESIGN METHODOLOGY

- Design
- Review
- Negotiate
- Revise
- Redesign
- Repeat until out of resources

INFORMAL DESIGN REVIEWS

Consultation with equipment suppliers

The people who sell unit operations and other process kit usually have a very deep knowledge of its practical characteristics, and those of competing products. Obviously they would like to sell you their kit, but they will scarcely ever lie in order to achieve this. See a good few of them, and you can learn how to play a game which will allow you to incorporate their detailed practical knowledge into your designs.

Consultation with electrical/software partners

Sometimes you will have in-house electrical or software engineers, but nowadays there will normally be an external electrical installer, motor control center (MCC) supplier and software designer. These may all be under one roof, or there may be combinations. If you don't have in-house specialists to back you up, combinations are better, but as with all in engineering, the less you know the more you pay.

Electrical and software components of the job are very significant, and are perhaps the single biggest opportunity for cost overruns at installation and commissioning stage, so there is a potential liability to manage. There are also big opportunities for cost savings if a well-integrated and controlled design can be devised.

Consultation with civils/buildings partners

As with electrics, civils and buildings are often not designed in-house. Civil engineering companies often work on very small margins, and may consequently have a somewhat inflexible approach to contract documentation.

They are far more likely to employ quantity surveyors (QSs) than other disciplines. QSs are a kind of engineering accountant-cum-lawyer, and are not well loved by engineers. They are characterized in an old joke as the "people who go in after the war is lost and bayonet the wounded."

These companies are also much more likely to sue partners if things do not go well in construction than other disciplines. Experienced engineers are consequently generally very cautious in their dealings with civil partners, though design and costing are normally separate parts of the operation for civil contractors and consultants.

There is, however, potential for both good savings and, more importantly, good control of potential construction stage cost overruns, if the civil aspects of design are well integrated and defined.

The things you learn in these discussions can also alter the starting point of your future designs in such a way as to give them better cross-discipline integration.

Consultation with peers/more senior engineers

Some people like to keep things to themselves, and some need a sounding board to develop ideas. I am in the second category, so I have learned a lot from others during these interactions. The other party does not, however, always have to be a more experienced engineer. Sometimes you just need to get an idea out there and play with it to see its strengths and weaknesses. Sometimes it needs a fresh pair of eyes to see things which an idea's author cannot.

Unless you are in the fortunate position of being allocated a personal mentor, senior engineers may often not have a great deal of time to talk to you, but few will refuse to help you out with a knotty problem.

Peers and near-peers will probably be the people you spend most time discussing things with and, while they can be useful, you should bear in mind that they are more likely to suggest investigating blind alleys, or using inappropriate design techniques than old hands.

FORMAL DESIGN REVIEWS

Interdisciplinary design review

The point of design reviews is to make sure that the design is reasonably optimal in the opinion of more senior engineers.

Designs need to balance the needs of (at a minimum) the process, mechanical, civil, and electrical engineering disciplines. Considering installation and commissioning issues is also mandatory.

The design review will therefore bring together the senior engineers of each discipline within the company, and will focus on the design drawings.

The atmosphere of such meetings is normally reasonably friendly, but engineers can be quite challenging. There may also be internal company political issues at play. Strong chairmanship and negotiation skills are a requirement if such meetings are to work well.

Beginners will learn a lot in such meetings, and should miss no opportunity to attend one, even though the first few for their own designs might include some learning experiences they might not entirely enjoy.

Value engineering review

Value engineering reviews have many similar characteristics to the design reviews of the last section, though their focus on cost and value will attract more management types, sales and marketing people, and so on.

As the name suggests, they are an attempt to get the price right—usually aiming for a downward adjustment.

These are sometimes conducted in the presence of client representatives, which can on occasion result in a rethink from scratch of some constraint on the design which the client had not realized the full implications of.

Safety engineering review

This is most often a formal approach such as a HAZASS or HAZOP, but it is culturally very similar to the last two types of review.

QUALITY ASSURANCE AND DOCUMENT CONTROL

Engineers ... are not superhuman. They make mistakes in their assumptions, in their calculations, in their conclusions. That they make mistakes is forgivable; that they catch them is imperative. Thus it is the essence of modern engineering not only to be able to check one's own work but also to have one's work checked and to be able to check the work of others.

Henry Petroski

Control of the design process and design documentation is incredibly important in professional practice. There are strong negative safety implications of poor control of the design process.

Engineering documents are, for example, always marked with revision numbers and dates. If the document is changed in any way, it is given a new dated revision number. This prevents a lot of potential confusion.

Once a design is at the point where the major design difficulties have been resolved (after usually 2—3 revisions), revision zero will normally be issued. Any remaining issues on the P&ID may be highlighted within an irregular outline and marked HOLD. Once this is done, the design is considered sufficiently complete for the drawings to be used as the basis for commencement of design and procurement.

Once it has been decided that the design is sufficiently complete to issue for construction, the design will be frozen, which is to say that no significant change will be allowed.

So document change control is very important—it is usual to specify the engineering discipline and degree of seniority required to change given documentation, and for there to be checking of all changes by a second engineer of an appropriate discipline with a specified level of seniority.

The ISO 9000 series European Standard (derived from BS5750, a British Standard for quality assurance in produce design and manufacture) gives a very widely used formal methodology for the control and audit of all processes. The ISO 9000 series is now applied away from product design and manufacturing, and there are less formal approaches used everywhere in engineering which cover more or less the same ground.

ISO 9000 requires that you have documented procedures for:

- Control of Documents
- Control of Records
- Internal Audits
- Control of Nonconforming Product
- Corrective Action
- Preventive Action

It also requires that you keep permanent records of the following:

- Management reviews
- Education, training, skills, and experience
- Evidence that processes and product or service meet requirements
- Review of customer requirements and any related actions
- Design and development including: inputs, reviews, verification, validation, and changes
- Results of supplier evaluations
- Traceability where it is an industry requirement
- Notification to customer of damaged or lost property
- Calibration
- Internal audit
- Product testing results
- Nonconforming product and actions taken
- Corrective action
- Preventive action

These are records you need to provide evidence of following your processes.

Professional process plant design is a very highly controlled and documented process, and your design decisions are recorded for future reference. It should be borne in mind that your decisions may even have to be defended in court long in the future.

Like many other engineers, I keep my own dated handwritten notebooks in which I record my decisions. They can come in handy if you ever have to appear in a court case in respect of a decision you made 10 years previously. Contemporaneous notes are admissible evidence, as well as an *aide-memoire*.

INFORMAL DATA EXCHANGE

Engineers love to talk about things which have gone wrong. This might seem like gossip, and there may be a component of that, but actually it spreads the knowledge of what doesn't work.

Engineering experience consists far more of knowing what doesn't work than knowing what does.

FURTHER READING

Anon, 2008. ISO 9001:2008 Quality Management systems—Requirements. International Standards Organization, Geneva.

Dearden, H.T., 2013. Professional Engineering Practice: Reflections on the Role of the Professional Engineer. Createspace.

CHAPTER 17

Beginner's Errors to Avoid

INTRODUCTION

It takes an engineer to undertake the training of an engineer and not, as often happens, a theoretical engineer who is clever on a blackboard with mathematical formulae but useless as far as production is concerned.

The Rev E.B. Evans

Academic myopia

Many of the errors which follow are often hard-wired into academic design techniques. In summary, best simply forget everything you were told in academia about process design. Those who taught you have almost certainly never designed a unit operation which has been built, let alone a whole plant.

Lack of consideration of needs of other disciplines

Real process plant designers have to take into consideration the needs and desires of several other engineering disciplines, most notably civil, mechanical, electrical, and software in that order. The idea that a chemical engineer can sit down and design a plant in glorious isolation comes only from the inadequacy of links between disciplines and professional practice in academia.

Lack of consideration of natural stages of design

It is one thing to consciously accelerate a program by rolling a couple of stages of design together. It is quite another to attempt to apply techniques intended for a parallel universe where these stages do not exist.

Excessive novelty

Academics progress in their careers by being radically innovative. Being novel is more important than being right to researchers who wish to be published, and many teach their students to value novelty too. Professional engineers are no more novel than absolutely necessary. Being right is far more important to us than being original.

Lack of attention to detail

"Block flow diagrams" are commonplace in university, and process flow diagrams (PFDs) are commonly the highest level of definition of plant interconnectedness. I have never used a block flow diagram in professional practice—I often go straight to

An Applied Guide to Process and Plant Design

255

piping and instrumentation diagrams (P&IDs). I only do PFDs if I need them to envisage mass flows, or a client asks for them.

This is symptomatic of the different levels of attention applied by professionals and theorists. Having taught process design to many university lecturers, I know that it is commonplace for the mediocre ones to think that all design problems below the level of mathematical theory are trivial. Only the exceptional ones are willing to throw themselves in to the point where they learn that the devil is in the detail.

There is a useful checklist in "Practical Process Engineering" to check for P&ID completeness which will draw a beginner's attention to frequently neglected issues, which I have reproduced in Appendix 4.

Lack of consideration of design envelope

Universities are graduating students who have never considered anything beyond static steady state design. Even Master's level "Advanced Chemical Engineering" modules use the simplified model so that they can spend longer on pinch analysis. This model is at best a simplified one used for the newest beginners—this is not how design is done.

The design envelope considers all relevant aspects of a specific proposed site in determining which approach is likely to be best. The regulatory environment, climate, price of land, skills of available operators and construction companies, reliability of power supply, risk of natural disasters, and proximity of people are often at least as important as the theoretical yield of a process chemistry.

There is no right answer to design. The right design for a less developed country will not be that for a more developed country. The right design for a client with a lot of experience with a particular process will differ from that for another client.

Lack of consideration of construction, commissioning, and nonsteady state operation

This is a subset of the above error. If your plant doesn't work during commissioning and maintenance it doesn't work at all. Get it right—consider all stages of the plant's life.

Principal maintenance activities to consider during design are isolation, release of pressure, draining, and making safe by purge or ventilation. Remember to allow for isolation of utilities as well. Purge lines should ideally be temporary to prevent back-flow contaminating the reservoir.

Isolation, block, double-block, and double-block-and-bleed valves are used for these duties, supplemented by spades, slip plates, and blinds.

If there is going to be hot maintenance (while the rest of the plant is working), the layout must be suitable for this. Isolation of vessels must not isolate them from the pressure relief system, though the possibility of connection to process via this route needs to be considered.

Parallel and series installation

Beginners seem to have little feel for the differences between parallel and series duplicate installations, and when each is appropriate. They may, for example, think that the headlosses of units in parallel are additive (they are not). The heuristic is: pumps in series—add heads, pumps in parallel—add flows.

Lack of redundancy for key plant items

Standby capacity can be something of a mystery to beginners. Generally speaking, if an instrument or item of rotating machinery is crucial to plant operation, you need at least one full size standby unit. For economic or other practical reasons you may alternatively choose to have three 50% duty units instead of two 100% units (duty/assist/standby versus duty/standby).

If the item is really crucial, you might want to make sure that a common cause failure of the units cannot happen. So you might specify, at a minimum, separate cabling all the way back to the motor control center (MCC) for electrically driven items, or follow the example of the oil and gas industry in having steam-powered backups for crucial electrically driven pumps.

It is not only unit operations which may need backup; utility failure can lead to hazardous situations. We need to assess the required reliability and include standby as required early on. Electrical power to crucial items usually requires twin feeds and/or generator/battery backup. Note that generator/battery reliability and the yielded length of trouble-free operation costs money, and that generator supplied power is not necessarily as "clean" as mains power. It may be that "load shedding" needs to be specified, such that only the most essential plant processes have power backup.

Emergency cooling or reactor dumping to quench tank may be used to bring the plant to a safe and reliable stop rather than provide for continued operation in the event of utility failure.

Lack of consideration of processes away from core process stream

Assume nothing. If the client has not told you that water, air, electricity, effluent treatment, odor control, and so on are available free of charge and suitably rated for your process, assume you need to provide them. If the client has told you they are, check that you are happy with what is offered. Mark your P&ID with termination points making clear where your scope begins and ends.

Storage often takes up more space, and presents more hazards than the main process. Note that a few big tanks are probably cheaper than lots of small ones but the consequences of failure are greater.

Designers need to consider emergency releases from the plant, allowing fenced off out of bounds sterile areas for flares and vents, and adequate scrubbers for toxics.

We need to make sure drainage systems are adequately designed. We need to avoid pits which might collect heavier-than-air flammable vapors if leaks could produce them, unless these are specifically designed impounding basins to allow leaking flammable vapors to burn off without damaging other equipment.

Tanks should be contained in bunds with 110% of the largest tank volume being the usual minimum allowance. We need to consider precipitation/firewater drainage requirements when designing bunds. They may need covering or additional capacity. Access (which does not breach the bund) to any equipment inside the bund also needs to be provided.

Lack of consideration of price implications of choices

The best university level pricing techniques are greatly inferior to professional practice, and many things considered perfectly acceptable in academia are considered woefully inadequate in professional practice. Making choices between technologies or configurations without pricing them as well as you possibly can at that stage of design is unprofessional. Don't do it.

Academic "HAZOP"

There is a thing called HAZOP (or sometimes CHAZOP) by many in academia which consists of reviewing a PFD using a version of the HAZOP procedure to generate the required control loops. This is neither HAZOP nor CHAZOP, and this is not how we determine how to instrument and control our plants. If you don't know how to do the basic control of your plant, look at Chapter 13, and/or ask an experienced engineer. You'll only be doing it their way come the design review anyway.

Uncritical use of online resources

The internet is full of all kinds of potentially useful resources. I once asked one of my students the source of some pricing data, and he said "Chinese Websites." Without getting as obsessed with proper referencing as my research colleagues, we do need to put a little more thought into the reliability of any internet resources we use.

There are all kinds of online calculators for sizing pipework, equipment, and so on. There are sites which advertise chemicals and equipment for sale. Some of these are good (such as, e.g., lmnoeng for fluid flow calculations), and some are very misleading.

Professional engineers do not offer printouts of stuff from the internet as a substitute for their own calculations based in reliable information. I have, however, been known to use lmnoeng and the like, as a quick check that I have any novel hydraulic calculations about right if there isn't a second engineer available to check me. I don't assume I have it wrong if the website disagrees, but if it agrees with me, I am happy to assume my calculations are about right.

Professional judgment is what engineers get paid for. Don't do anything without exercising it.

LACK OF EQUIPMENT KNOWLEDGE

Lack of knowledge of pump types and characteristics

There are two main kinds of pumps (rotodynamic and positive displacement), whose characteristics were explained in an earlier chapter. There are many subtypes of pumps which differ from each other in nontrivial ways.

The type of pump selected affects the way the pump needs to be controlled and protected, the precision of pumping, the suitability for a given fluid in terms of its viscosity and solids content, the power utilization for a given duty, the maintenance requirements, and so on.

Until beginner designers apply the knowledge of pump characteristics outlined in the Tables 10.3 and 10.4, they will consistently make schoolboy errors in the selection of pumps and surrounding systems.

Attempting to control positive displacement pump output with a valve

This is one of the knock-on effects of the broad class of errors covered in the last section, and one of the commonest errors of those with little design experience (which I have seen in supposedly bright young engineers applying for chartership).

Do not attempt to control the output of a positive displacement pump with an in-line throttling valve—this does not work and will damage the pump.

Multiple pumps per line

There is nothing at all wrong with having multiple pumps in parallel as an assist or standby arrangement, but pumps in series are usually an error. Pumps in series, especially multiple positive displacement pumps in series, are not a feature of professional designs.

Professionals know that we can get multiple stages of centrifugal pump in a single unit if we want higher delivery head, and if you can't get a pump to do your duty, you are probably looking at the wrong kind of pump.

We also know that multiple positive displacement pumps in series do not work, as we are throttling suction or delivery when they pump out of synch.

Lack of knowledge of valve types and characteristics

There are essentially three broad classes of valve duties, as outlined in the earlier section: isolating, on/off, and modulating control valves. Different industries use different valve types for these duties, but all industries have these requirements. All designs should reflect this understanding. Table 10.5 is intended to help beginners to understand more about what is available. Actuated valves should be considered as rotating machinery—if crucial to the process, standby capacity is required.

Lack of knowledge of actuator types

As outlined previously, there are three broad kinds of actuators. Note that pneumatic actuators require compressed air and additional control equipment (solenoid valves to control airlines) to function.

Throttled suctions

Don't try to control the output of a pump by throttling the suction, so as to avoid cavitation, among other things.

Use of actuated bypass valves

Back when I was in university it was common to control the output of a positive displacement pump with an actuated valve in a bypass to suction. It does at least sort of work, but times have moved on, and we use inverters now. I never used this technique to control centrifugal pumps even back then.

Use of control valves

I personally don't make a lot of use of in-line control valves for liquids nowadays at all. While this is personal preference, the greater power efficiency of inverters is a fact. In the United Kingdom there are tax breaks for using inverters because of this.

Multiple valves per line

In university, they might have taught you a clever way of using multiple valves in the same line with different lags to control flow based on multiple variables, but I would recommend that you don't try it in practice unless there is absolutely no other way of achieving your aim. KISS. Try to control flow in a line only once. In any case most of the multiple valves per line I see in beginner's designs are not sophisticated cascade control, they are simply errors.

Lack of tank drains and vents/other valves necessary for commissioning

A word of warning—don't upset the commissioning engineer. Commissioning engineers want to be able to drain tanks down in a reasonable time—say 30 min. Make it so. Air will need to come in to replace the fluid—be sure to include a vac/vent or other valve to allow this.

Think about the commissioning operation—additional valves may be needed to commission unit operations in isolation, or add services needed during commissioning. Put them in.

If you are unsure about what commissioning engineers need, ask them. If you do, exercise professional judgment and be prepared not to add absolutely everything they ask for. They are not employed to care about whether the company gets the job.

Lack of consideration of details of drainage systems
Badly designed drainage systems can be the cause of very serious problems—they can allow the build-up of hazardous material from leaking equipment though undersizing or lack of provision for removal of solids build-up, allow incompatible materials to mix, carry toxic gases, fire or explosions from one section of the plant to another. They are nontrivial.

Lack of sample points
Commissioning engineers will also berate you for omitting the valves they need to take samples while commissioning. As with all the things which upset commissioning engineers, operating staff won't thank you either. Think about where you will need to take a sample to test whether a unit operation is working. Put a sampling valve in there, or a more complex arrangement if containment is an issue.

Lack of isolation valves
Every unit operation needs at least one isolation valve on every inlet and outlet. Include it.

Lack of safety valves
Nonreturn valves, pressure relief valves, pressure sustaining valves, etc.: if you haven't included them you haven't really considered all that can happen on the plant. More experienced engineers will hopefully add what you have omitted, but why not save them the trouble?

Lack of redundancy for key valves
Key actuated valves may well need actuated standby valves, and all actuated valves on units which are not themselves entirely duplicated are likely to need a bypass with isolation and a manual standby control valve for maintenance in service.

LACK OF KNOWLEDGE OF MANY TYPES OF UNIT OPERATIONS

The law of the hammer (if all you have is a hammer, everything looks like a nail) operates if you know too little about your options.

Universities seem to concentrate on a small selection of unit operations important in the petrochemical industry. All chemical engineering students tend to see scrubbing, stripping, distillation, and drying several times during their course. The other 99% of unit operations are a mystery to them.

This book attempts in some of the tables in Chapters 10 and 11 to address the issue of lack of knowledge of separation processes and so on. More generally, new designers need to discuss the things they are doing with more experienced engineers, so that they at least get a chance to know what they do not know.

LACK OF KNOWLEDGE OF MANY MATERIALS OF CONSTRUCTION

My students used to know about two materials of construction, which they used for everything—carbon steel and stainless steel (they usually weren't sure which grade). There is rather more choice than that, as Table 10.1 shows.

LACK OF UTILITIES

Make sure all utilities are included at earliest stages, for example cooling water, nitrogen, and refrigeration as well as steam, process water, electricity, and compressed air.

If you are handling highly flammable materials, one way to make them safe is to exclude oxygen from vessel headspaces with inert gas. Nitrogen is cheapest, though sometimes more exotic gases are required. You need to make and/or store this on site.

LAYOUT

2D layout

Beginners to plant layout consistently fail to think in three dimensions—they lay pipework and plant out on the floor in plain view in a way which renders it a dense series of trip hazards, instead of fixing it to the walls or grouping in pipe racks and bridges like real engineers.

Lack of room and equipment for commissioning and maintenance

This is the layout version of steady state design myopia. Detailed consideration needs to be given by the designer to how the plant will be accessed during commissioning and maintenance activities. The safety implications of this make it a high priority.

Lack of control rooms and MCCs

As mentioned previously, new designers may be unaware that we need MCCs to control the plant, and that we normally put these in a control room. The control room size needs to take into account the direction from which the panel is accessed for maintenance, the direction the cables come in from, and be big enough for safe access with the MCC doors open. There will also normally be a table with a PC on for system control and data acquisition (SCADA), room for filing cabinets for paperwork, etc.

PROCESS CONTROL

Lack of redundancy for key instruments and safety switches

Beginners tend to miss out key instruments entirely, and slightly more experienced engineers can fail to allow for standby capacity for safety or process critical instrumentation. Such standby provision needs to be balanced against the need for simplicity.

Lack of isolation for instruments

Instruments need maintenance and replacement. Unless you only propose to do this with the entire plant shut down and drained, isolation valves are recommended to allow removal and replacement when the plant is running.

Measuring things because you can, rather than because you need to

Don't measure things you can't control. It will only cost you money, and it might upset you needlessly.

Alarm overload

Consider the number of alarms you are generating—don't overload operators with more alarms than they can take in. This will make the plant less, rather than more, safe.

P&ID notation

We mostly control plants with programmable logic controllers (PLCs) or distributed control system (DCS) systems nowadays, so P&IDs should usually not show control loops as if they were wall-mounted proportional, integral, differential (PID) controllers as shown in Figure 17.1:

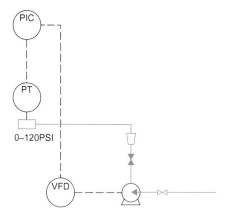

Figure 17.1 P&ID notation.

FURTHER READING

Sandler, H.J., Luckiewicz, E.T., 1987. Practical Process Engineering: A Working Approach to Plant Design. McGraw-Hill, New York, NY.

CHAPTER 18

Design Optimization

INTRODUCTION

Premature optimization is the root of all evil.

Donald Knuth

Contrary to popular academic opinion, availability of accurate information means that design optimization is most usually and always best applied after a plant has been built, commissioned, and operated for a while.

Design optimization tools such as modeling, simulation, and pinch technology are therefore poorly suited to use during the plant design process, with a few notable exceptions. Their use in academia is almost always misuse grounded in a lack of understanding of the constraints of professional practice.

MATCHING DESIGN RIGOR WITH STAGE OF DESIGN

The main thing that those who apply academic process optimization techniques to process plant design fail to understand is that the iterative nature of real design processes means that there is already a complex design optimization process going on.

Professional engineers, however, understand that each stage of design has its own natural resolution. There is no practical point in applying a technique with a resolution finer than the model it is being applied to.

The second thing which the academic approach fails to address is that you cannot meaningfully optimize a model which has not been verified by input of real-world data.

Like microscopes, all design techniques have what we might call a limit of resolution. Microscopic resolution allows us to distinguish accurately between two lines. A microscope with insufficient resolution for the task to which we put it may give the appearance of two lines where there is really only one, or one line where there are actually two.

Similarly, a design technique with insufficient resolution may make two options seem equal where one is actually better, equal options significantly different, or even the better one worse. Misuse of process optimization tools for design is to me akin to what is known in microscopy as "empty magnification," where you make an image look bigger, but it actually holds no additional information.

Process integration

What is known in academic circles as "process integration" is not design integration. It probably isn't even really process integration. The professional process designer's "process integration" balances a number of mutually dependent considerations. The design needs to be safe, robust, and cost-effective, but safety and robustness do not come for free. A balance has to be struck.

As piping and instrumentation diagram (P&ID), general arrangement (GA), process flow diagram (PFD), process, and hydraulic calculations are developed, many choices have to be made about the broad outlines of plant layout, degree of redundancy of equipment, and basic approaches to safety.

Potential hazards have to be identified, quantified, eliminated, or controlled. Materials and equipment have to be specified. While doing this, installation, commissioning, maintenance, and nonsteady state operation of the plant have to be considered. Past experience with other similar plants needs to be incorporated.

The process designer does not do this in a vacuum—they need to integrate the requirements of and insights from other disciplines. Optimizing a few aspects of the process, or even the whole process chemistry, is not optimizing the overall plant design. It may actually be making it less optimal.

What is often meant by process integration in academia is use of a mathematical analysis of a system using one of what is now a wide range of mathematical, graphical, or computer-based tools, originally developed for beginners.

The problem these tools solve is one of handling a multiplicity of possible solutions. It isn't so much that there are an infinite number of possible solutions to the question, each of which has a number of subtly different implications, as that there are a great number of permutations to winnow for the best value of a single numerical selection criterion.

The tools can perform this winnowing process for us, but the fact that there is essentially one right answer, and a computer can find it better than a person, tells us that this isn't really engineering, and the problem is essentially trivial.

These tools may have some limited use in the final stages of designs which use a lot of energy, and have clear possibilities for substantial recovery of that energy. They may also be of use in identifying possible improvements to existing processes.

Starting a design from heat integration of a process at steady state without consideration of cost or other implications is trying to fit a job to a tool rather than the reverse.

Another buzzword in academia is "process intensification." Professional engineers make processes as intense as they practically can, but no more so, to paraphrase Einstein.

INDICATORS OF A NEED TO INTEGRATE DESIGN

Professional process plant designers always integrate their designs (it's the most important aspect of process design) though they are integrating different aspects than those using the term in academic contexts. There are, however, certain contexts where they address the same issues as the theorists, most notably where there are likely to be big cost implications.

High utilities usage/waste

Engineers will always be concerned that their design might be less than optimal if they see that a design has high operational costs due to high utilities usage.

I worked for some years for a UK government scheme called Envirowise in which we visited process plants and factories to audit resource usage. What was clear to me after doing a hundred or so of these visits is that there are a number of areas where such wastage is commonplace. In fact Envirowise eventually gave us a table for the best places to start looking, which I have reproduced later in this chapter.

The use of this table is far more economical in terms of designer resources than carrying out a mathematical network analysis such as pinch. We have to conserve our resources too.

High feedstock use/waste

Feedstock costs money, and the waste streams generated by wasted feedstock often cost money to dispose of. This does not imply that the optimal feedstock conversion rate is 100%. Each incremental increase in conversion usually costs more than the last similarly sized increment.

Back when I worked for Envirowise we were encouraged to nudge people in a direction toward zero waste, but there is almost always a point on that road beyond which economic viability becomes questionable. It is, however, true to say that uncritical acceptance of traditional levels of resource inefficiency is unlikely to yield optimal design either.

A bit of analysis of resource usage is almost always informative and worthwhile for operating companies. Designers should, however, already be taking these issues into consideration using the standard combination of mass balance, appropriately accurate costing and sensitivity analysis.

HOW TO INTEGRATE DESIGN

Since the mathematicians have invaded the theory of relativity, I do not understand it myself anymore.

Einstein

Professional designers are employed to produce integrated designs, but the things they balance are practical things like cost, safety, and robustness, process controllability and operability, and the conflicting demands of the various disciplines and stakeholders involved.

They never optimize designs for a single variable or small number of variables, which is why process plant design is more like playing chess than doing logic puzzles. In fact it is far harder than chess, which is why people can still outdesign computers, but computers are now the world's best chess players (the fact that computers can do it better than people shows that chess is in fact a very complex task, rather than a problem solving exercise). There may be no right answer to a design exercise, but there are better and worse answers, and better and worse players.

Those who think that process plant design is or will ever be a form of applied mathematics simply do not understand the nature of design. They have simplified an activity to a level where its point has been lost.

Intuitive method

Much professional engineering knowledge is qualitative or semiquantitative. Envirowise produced a number of graphs and tables to facilitate increased resource efficiency of existing processes which designers can use to inform their design process, which are reproduced in Tables 18.1—18.3 and Figures 18.1 and 18.2.

There are also useful tables on waste heat recovery in "Practical Process Engineering" (see "Further Reading").

These tables and charts are as useful to a beginning plant designer as an operating company, as they show us where to look for improvement to our designs, and the rough cost/benefit profile.

Table 18.1 Resource efficiency measures for process plant
Cost-effective water saving devices and practices for industrial sites: process plant[a]

Item/Application	Description/Purpose	Equipment/Technique	Applicability	Other benefits	Other considerations	Potential cost	Potential payback
Liquid ring vacuum pumps	Reuse of sealing water after treatment	Tanks/pumps/separators/cooling	Widespread	Energy savings from cooling seal water	Seal water: temperature and quality control	H	M
	Eliminate water use	Mechanical vacuum pumps	Widespread		Liquid trap	H	M
Cooling towers	Automatic blowdown—operation at maximum acceptable total dissolved solids (TDS) level	Conductivity-base control	Widespread	Reduced chemical use		M	S
	Cooling load reduction—minimize evaporation and blowdown		Widespread	Reduced chemical use		M	M
	Alternative cooling processes to avoid evaporation of water:						
		(i) Air blast	High cooled water temperature (>40°C)		Monitoring requirements	H	M–L
		(ii) Heat exchangers	Widespread	Waste heat could be used elsewhere		H	M
Heat exchangers	Water reuse through closed loop system	Tanks/pumps/heating source/cooling source	Widespread	Heat sink/cooling tower/water quality		H	M–L
Hydraulic power packs	Optimize water use by varying water flow depending on oil temperature	Bulb-and-capillary operate control valves	Widespread		Essential cooling requirement	L–M	S
	Reuse after cooling—through closed loop system	Tanks/pumps/cooling source	Large installations		Cooling tower/water quality	H	L

Potential costs and paybacks are for guidance only. Actual costs and paybacks will vary due to project-specific details.
[a]Risk assessment required; Potential cost: L = low (minor alterations) (£0 to a few £100s); M = medium (a few £100s to a year); H = high (extensive alterations or new plant required) (many £1000s)
Potential payback: S = short (a few months); M = medium (less than a year); L = long (over a year).
Copyright material reproduced courtesy of WRAP/Envirowise.

Table 18.2 Resource efficiency measures for cleaning and washdown

Cost-effective water saving devices and practices for industrial sites: cleaning and washdown[a]

Item/ Application	Description/ Purpose	Equipment/ Technique	Applicability	Other benefits	Other considerations	Potential cost	
Pressure control/flow restriction	Reducing instantaneous flow at point of use	Valves, orifices, pressure-reducing valves	Variable or intermittent supply, pressure or demand			L	S
Countercurrent rinsing	Reuse of rinse water	Tanks	Multistage unit processes		Water quality requirements	M	L
Spray/jets	Appropriate application of water	Nozzles	Widespread	Improved cleaning		L–M	S–M
		Spray nozzles	Widespread	Improved cleaning	Spray mist drift	L–M	S–M
		High pressure spray packages	Washing processes	Improved cleaning	Power consumption	M–H	S–M
Automatic supply shutoff	Use of water only when needed	Solenoid valves in pipelines	Small bore pipes		Essential water requirement	L–M	M
		Actuated valves in pipelines	Large bore pipes		Essential water requirement	M	M
		Jets/ spray guns on hoses	Widespread	More efficient application	Theft of spray guns	L	S
Reuse of wash water	Reuse of wash water in other areas	Tanks/pumps	Widespread		Cross-contamination/ water quality control	M	S–M
Scrapers/ squeegees/ brushes	Sweeping up of slurries	Dry cleaning methods	Large areas	Possible reuse of materials	Dry collection systems	L	S
Cleaning-in-place (CIP) technology	Countercurrent reuse of rinse water with multiple reuse of chemical cleaners	Proprietary plant	Processes with frequent cleaning	Hygienic plant/ minimal downtime for cleaning	Water quality requirements	H	S–M

(Continued)

Cost-effective water saving devices and practices for industrial sites: cleaning and washdown[a]

Item/ Application	Description/ Purpose	Equipment/ Technique	Applicability	Other benefits	Other considerations	Potential cost	
Recycle after treatment	Treatment of wastewater to an acceptable standard for reuse	Filtration/ sedimentation	Coarse solids removal/phase separation		Waste disposal and water quality	M	M–L
		Centrifugation/ flotation	High quality solids removal/phase separation		Waste disposal and water quality	H	M–L
		Biological treatment	Removal of dissolved biodegradable solids		Waste disposal and water quality	H	M–L
		Ion exchange	Removal of dissolved contaminants		Waste disposal and water quality	H	M–L
		Distillation/ stripping	Solvent recovery	By-product	Waste disposal and water quality	H	M–L
		Absorption/ adsorption	High quality treatment, solvent recovery, removal of toxic substances, color, etc.		Disposal of spent absorbent	H	M–L

Potential costs and paybacks are for guidance only. Actual costs and paybacks will vary due to project-specific details.
Potential cost; Potential cost: L = low (minor alterations) (£0 to a few £100s); M = medium (a few £100s to a few £1000s); H = high (extensive alterations or new plant required)

[a]Risk assessment required; Potential cost: L = low (minor alterations) (£0 to a few £100s); M = medium (a few £100s to a few £1000s); H = high (extensive alterations or new plant required) (many £1000s); Potential payback: S = short (a few months); M = medium (less than a year); L = long (over a year).

Copyright material reproduced courtesy of WRAP/Envirowise.

Table 18.3 Typical water savings

Water saving initiative per project	Typical reduction per site
Commercial applications	
Toilets, men's toilets, showers, and taps	40% (combined)
Industrial applications	
Closed loop recycle	90%
Closed loop recycle with treatment	60%
Automatic shutoff	15%
Countercurrent rinsing	40%
Spray/jet upgrades	20%
Reuse of wash water	50%
Scrapers	30%
Cleaning-in-place (CIP)	60%
Pressure reduction	See Fig 18.1
Cooling tower heat load reduction	See Fig 18.2

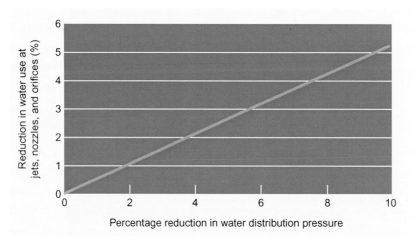

Figure 18.1 Effect of pressure reduction on water use at jets, nozzles, and orifices. *Copyright image reproduced courtesy of WRAP/Envirowise.*

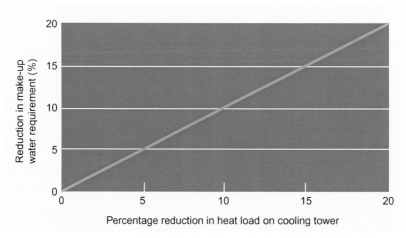

Figure 18.2 Effect of heat load reduction on make-up water requirement for a cooling tower. *Copyright image reproduced courtesy of WRAP/Envirowise.*

Formal methods

Pinch analysis

Originally intended for optimizing heat recovery, pinch analysis has also been applied to analysis of mass flows, including water flows. It is neither novel, nor much to do with professional design, but academics love to apply it to design. I learned it in university, and not only did I never use it, but I never once heard it mentioned until I entered academia again 20 years later.

An outline of the most commonly used method for the production of water purity profiles with a fixed flowrate for a single contaminant is as follows:

1. Draw a graph of flow rate versus concentration for all sources and sinks of water on a plant, where x-axis is flow, ascending from zero, and y-axis is concentration of contaminant, descending from zero.
2. Start on the left plotting flow rate/concentration pairs for potential sources of water, in increasing order of purity (dirtiest on the left).
3. Join the points to form a stepped "curve."
4. Next plot flow rate/concentration pairs for potential sinks and join the points to form a second "curve."
5. Move the sinks curve to the left until the curves just touch.
6. Where the two curves touch is the pinch point.
7. Where these two curves overlap represents the scope for water reuse.
8. Area to the left of any overlap represents wastewater generation.
9. Area to the right of any overlap represents clean water use.

Next, we can use sensitivity plots of potential cost saving versus concentration change for multiple contaminants, to identify areas where variation in maximum allowable inlet and outlet concentrations would yield the greatest savings.

We can consider four possible levels of investigation of the water reuse possibilities while carrying out our pinch analysis.

The cheapest and quickest analysis assumes that all sinks are presently at their maximum allowable inlet concentrations. This level of analysis will identify some cheap modifications which will yield small benefits.

Next we might consider the possibility of increasing the maximum allowable inlet concentrations in those areas where our sensitivity analysis has indicated that large savings might be available. There are technical limits on how far these concentrations can be increased without causing corrosion or other problems, which should be established and considered. Greater savings are likely to be obtained by this more in-depth analysis than for the simpler one above, and the costs of identified modifications are likely to be quite low.

A more rigorous analysis still considers the possibility of reuse after regeneration by water treatment technologies of a number of key streams. This can involve significant expenditure on water treatment plant.

Similarly, we might consider distributed effluent treatment techniques—rather than mixing all effluents together prior to treatment, we can consider treating or partially treating wastewater streams individually. It is claimed that this technique can offer improved contaminant removal efficiency at reduced cost.

Note that we need meaningful data on water quality and quantity produced and used to even start this process. People operating a real plant can obtain such data, but it is not available to plant designers. The same is true of all other pinch analysis techniques, which is the first reason why real process plant design engineers don't use it—they can't.

WHEN AND HOW NOT TO INTEGRATE DESIGN

Pinch analysis has no part in professional design exercises, and its inclusion in the design process would not merely be an extraordinary waste of time. Its use necessarily leads to suboptimal design. Amusing as it might be as an academic exercise to play with novel but unrealistic approaches, the downside of pinch analysis being the starting point for plant design is so obvious to professional engineers than no one ever does it.

WHERE'S THE HARM? THE DOWNSIDE OF ACADEMIC "PROCESS INTEGRATION"
Capital cost of "integrated" plants

When applied to energy use, the marginal cost/benefit ratio of each additional heat exchanger is not really considered by many of the techniques used by pinch analysis enthusiasts. This means that they are likely to be recovering heat at a greater cost than it can be purchased for. Real engineers don't do that.

It can be seen from the Envirowise table that the cost of additional heat exchangers is rated as "high," but that the potential for payback is only medium–high. Every additional increase in energy recovery needs to make financial sense. These very significant costs are not taken into consideration by process integration techniques as applied to design.

Commissioning "integrated" plants

The integration of heat exchanger networks means that the plant produced would be normally operated in a highly interdependent manner, and (theoretically at least), with reduced energy inputs.

The process commissioning engineer will have to get such a plant from an initial condition, in which even single systems are not locally balanced with respect to their flows and control loops, to one in which there are extensive cross system mass and energy balances. This is going to take additional time and other resources in the most resource-pressured part of the job.

If the designer has not included heating and cooling services capable of providing the full heat loads, ignoring integration, the commissioning engineer will need to make provision for them.

Small incremental savings in energy recovery may easily save less over the plant lifetime than the extra commissioning time and resources cost. These very significant costs are not taken into consideration by process integration techniques.

Maintaining "integrated" plants

Maintenance of integrated plants will carry the same problems as commissioning, and all the additional equipment will have its own additional maintenance requirements. These very significant costs are not taken into consideration by process integration techniques.

FURTHER READING

Anon, 2005. GG523 Cost-effective Water Saving Devices and Practices—For Industrial Sites. Envirowise.

Sandler, H.J, Luckiewicz, E.T., 1987. Practical Process Engineering: A Working Approach to Plant Design. McGraw-Hill, New York, NY.

CHAPTER 19

Developing Your Own Design Style

INTRODUCTION

There is a great deal more to engineering than the stuff they teach you in university, but I'm not talking about the ethics and embedded humanities modules which sometimes get shoehorned into curricula. There are even a few useful books on the subject.

THE ART OF ENGINEERING

Engineering design is not applied science. It is an art, learned and refined through practice. Engineers use all that they are in the practice of their profession. Intelligence and knowledge which are neither scientific nor mathematical are of crucial importance.

When I teach design to mature postgraduate students, it can seem as if they are more creative than the undergraduates, as they come up with a lot more ideas than the students with no industrial experience (but higher entry qualifications). I do not think that this is pure creativity. I think it is that they have more life experience to be creative with.

Judgment, intuition, and the knowledge and experience which teach us what doesn't work and enables us to reason by analogy all take time to develop.

Back when I was learning to teach I wrote a blog reflecting on my experiences, from which an excerpt follows:

> I went to see a client today. He had a problem, and had changed five things which might have caused it, as well as several others which might not (though he didn't understand that). I knew which two were the causes in ten minutes.
>
> I think to myself: 1. Engineering is easy (for engineers). 2. How did I learn how to do that? How can I teach others to do it? I could try this example as a case study, and see how hard it is to people earlier in their training. It seems to me at present that more so than amassing factual knowledge, it's to do with acquiring the engineer's perspective. Whilst it may have its limitations, on its home turf, it can cut through confusion, obfuscation, and misunderstanding in a flash.
>
> There is, however, no substitute for having a firm grasp of practical math and physical science. Theory underpins practice, and is available for verification of intuitive understandings. An experienced professional is not necessarily doing math and science in their heads when troubleshooting a problem. It is more like pattern-matching "Oh, yes, this reminds me of that time when..." and not necessarily even the words, just seeing into the problem, pruning the tree of possibilities. This involves people, and discourse (though engineers do not call it that).
>
> I spent far more time talking to the maintenance technician yesterday than I did looking at the machinery. In talking to him, I have to get him to talk freely, so that he tells me what he thinks has been happening. I have to assume that he will see what he expects to see, and

make sure I trust nothing of what he says which I have not verified personally. I look for areas where what he says is self-contradictory, and explore those areas with him in a way which does not make him feel I am trying to trap him into admitting he screwed up, or rubbish his pet theory.

I am, however, quite ruthless in making sure that I get to the bottom of what is happening to my own satisfaction. I'm going to find out what is wrong, and I'm going to fix it. How much I tell his boss is, however, negotiable. He knows how I work, since we have been interacting for a year or so, and we conduct an unspoken negotiation between us.

I am commercially interested in extending and upgrading the plant, but am constrained by professionalism not to milk the client. He is paid to maintain the plant, but would like it to be as automated and reliable as possible to make his job as easy as possible. He is, however, also paid to minimize costs, consistent with meeting the required effluent quality.

Between us we come up with a plan which makes us both look good, him cost conscious to his boss, and the client actively addressing the effluent failures from the point of view of the authorities. It also has a bit of what both he and I want, which is to pay me to make the plant better from the point of view of the maintenance staff as well as the other parties.

I'll also put some nice things in my report to his boss about the build quality of the modifications he has made and underplay their contribution to the problems. He in return will not mistreat the plant in between visits and blame it on my design. All of this is unspoken, but I know it is going on, and I think he does too.

Away from science and engineering, they might think chemical engineering is all about numbers and chemicals, but it seems likely to me that professional practice has less of this than academia. Working with other people's fears and desires, their wishful thinking and self-deception, strengths and shortcomings (as well as our own) is a crucial part of the job-but that doesn't mean a psychologist could do it.

THE PHILOSOPHY OF ENGINEERING

Ultimately the philosophy of engineering is as useful to engineers as philosophy of science is to scientists. "*Philosophy of science is about as useful to scientists as ornithology is to birds*" as Richard Feynemann is supposed to have said.

You might, however, have noticed that I have read quite a few books on the philosophy of engineering for someone who dismisses philosophy so readily, but I have only recommended books on the philosophy of engineering written by engineers. When I came to have to teach others what I had learned, but no one had taught me, I needed to figure out what I knew and how I knew it to be true.

It seemed before I read these books that figuring out what I knew about engineering and how I knew it was the subject of philosophy of engineering. However, it turned out that the subject of philosophy of engineering was philosophy, not engineering. I did find out one useful thing though, which was that much of what I knew was far from scientific.

Until this point I still thought that engineering was applied science. Anyone that has actually practiced the profession and had time to reflect upon it will see that it is far more art than science (though those who have never practiced, and those who have never reflected upon their practice might disagree).

So philosophy of engineering might be useful after all, as a way to get nonpractitioners to understand in an abstract intellectual way that engineering's foundation is not science and math, but *praxis*, the process of design.

This intellectual knowledge would not, however, be the fruit of praxis itself, and an academic informed by philosophy as to the nature of engineering would be a philosopher who understood what engineering was rather than an engineer. They might try to remember that if they write books on engineering.

THE LITERATURE OF ENGINEERING

I read a lot of books on process design in preparation for writing this book, and most of them were worthless. Books on design by people who had never designed anything (whether they are philosophers or any other kind of academic) are as accurate and informative as a braille list of rainbow types written by a person who had only ever heard them described.

I did not read those books to learn what process design is. The approach given in this book is not derived from the books I recommend, I have merely recommended those books which seemed to be informed by an understanding of what design is about.

Of far greater use to engineers than philosophizing are readable books which move our understanding of the nature of engineering forward. We can actually learn what doesn't work and why it doesn't more efficiently from books than from practice.

I would recommend, at a minimum, reading:

Trevor Kletz, An Engineers View of Human Error (or anything else by him).

Henry Petroski, To Engineer is Human (or almost anything else by him, though later books are not so good).
Harvey Dearden, Professional Engineering Practice: Reflections on the Role of the Professional Engineer.

If you are in a situation later in your career where you need to understand what you know without getting lost in abstraction, I would recommend:

Walter G. Vincenti, What Engineers Know and How They Know It.

Eugene S. Ferguson, Engineering and the Mind's Eye.
Donella H. Meadows, Thinking in Systems.
Billy Vaughn Koen, Discussion of the Method.

THE PRACTICE OF ENGINEERING

Accept no substitute, for there is none. Psychologists disagree with each other about the relationship between practice and mastery, but only because many of them aren't really scientists. They think it means that practice may be irrelevant to mastery if some never achieve mastery despite great amounts of practice, and some achieve mastery with very little (reported) practice. People lie, and anyone who teaches anything can see that some people have little aptitude for their subject.

Whatever your starting point, there is no substitute for persistence, as Calvin Coolidge said:

> *Nothing in the world can take the place of persistence. Talent will not; nothing is more common than unsuccessful men with talent. Genius will not; unrewarded genius is almost a proverb. Education will not; the world is full of educated derelicts. Persistence and determination alone are omnipotent. The slogan Press On! has solved and always will solve the problems of the human race.*

Some will find during practice that they are practicing the wrong thing, and change course, and there will always be an advantage to those with talent, but those who persevere will become Engineers.

Like Casey Ryback, I also cook, and I find practicing engineering is far more like cooking than it is like doing exam questions. If your educational assessment only consisted of completing exam questions you might find that what you are good at is exams rather than engineering.

PERSONAL SOTA

Koen calls the set of heuristics used by an individual their sota (for state of the art), and uses quasi-mathematical notation to show it as being that of an individual at a given time thus: sota $|_{\text{sean moran; 2014}}$ would be mine as of now. This may be a little gimmicky, but it does allow him to illustrate the relationships between individual sotas and best practice as the intersection on a Venn diagram of all sotas. So this book is a subset of sota $|_{\text{sean moran; 2014}}$, which I have gone to some effort to ensure corresponds with current best practice, sota $|_{\text{Eng; 2014}}$.

My sota depends upon my background, as does yours. Engineers are not generic—there are reasons why I am the kind of engineer I am (feel free to skip this bit if you think it self-indulgent).

All of my siblings are engineers. My father was not a professional engineer, but he was a pipefitter and mechanic who worked in engineering, specifically the operation and maintenance of power stations.

I have been hearing the stories engineers tell of what works and what does not along with explanations of the bits of kit being discussed all of my life. I also heard the stories maintenance staff tell about the folly and arrogance of green professional engineers who think book-learning alone is superior to experience and "technician level knowledge."

We didn't have a lot of money when I was a kid, and I had to learn how to mend my own stuff, as well as help my dad mend the car and do on. I also used to take apart any bits of machinery I found to see how they worked, and when I wanted a bike, I built more than one from discarded bits. I also built myself various types of computers. So I went into education already knowing quite a lot of useful things about the nature and culture of engineering, and with hundreds of hours of fiddling with engineered products.

A very un-PC technician got us all together when we covered process technology in my first degree, and said to us all "those of you who like fixing your own cars and bikes are going to do well in this course. Those of you (and I'm talking about the girls and Asians here) who don't aren't going to like it at all". He was wrong about women and people of Asian descent, but he was right about the link between a history of tinkering and a feel for engineering.

I am, by original education, an applied biologist. Biology was more about classification than deep understanding back when I started studying it. Molecular biology has elucidated the mechanisms behind much that was obscure back in the 1970s, but despite progress in systems biology, multicellular life is still very far from complete explanation. The subject is still mostly about the readily observed but ill-explained emergent properties of irreducibly complex systems.

The most important math in biology is still statistics, and often the nonparametric statistics of populations which do not meet the assumptions underlying the more commonly used kinds of statistics. Biologists understand that statistics is itself a set of heuristics, true only if their underlying assumptions are true. The most commonly used kind of statistics are not very robust, but are frequently used where their underlying assumptions are not true, rendering them at best meaningless. There is no such thing as 2.4 children.

Biology is an essentially qualitative field of study, whose objects are complex biochemical and physical processes controlled by homoeostasis, in which variables are regulated so that internal conditions remain stable and relatively constant. Despite this vagueness, we have for a very long time been able to make biology do things that we want it to.

So for me, a process plant is like a biologist's plant. I don't worry about whether I understand every aspect of the reductionist science of its subcomponents, or even whether it is possible to do so. They are unimportant to the job of engineering a plant which will give me the yield I want of product in a safe, reliable, and cost-effective way.

Then there is the element of chance and the opportunities available to us—I originally trained as a biochemical engineer, and ended up in water treatment plant design due to unexpected political resistance to biotechnology, and the cyclical nature of the water industry. My process plant design experience was mostly with contracting companies, and my job title was proposals engineer, rather than process engineer. I had to design whole plants, coordinate all disciplines, and care a great deal about cost, risk, footprint, and so on. I think this is why this book has a different scope from other professionals whose job title was or is process engineer.

Such books tend to be limited in scope to very detailed design of unit operations, because that is what process engineers are often asked to do. Engineers of other disciplines (notably mechanical) coordinate the design effort, and process engineers are just used to carry out the chemically bits the mechanical engineers cannot do. This is my working explanation for why process plant design books by process engineers are narrow, deep, and limited in scope, and the best book I could find on the subject was written by mechanical engineers.

Then there is the question of attitude—I started out overconfident, nearly won a job which I would have regretted winning, and became more cautious. I have seen overcautious engineers tread a reverse path. Experienced professional engineers still differ from each other in their risk aversion, but both overconfidence and its opposite are corrected by experience in those who stay in engineering. Other personality traits also often tend toward the middle way as a result of practice.

Your background and experience will be different, but practice will shape you into a particular kind of engineer. You will have a personal state of the art. You will find yourself seeing all kinds of problems away from professional life as soluble with the tools of engineering. You will come to understand that in life, as in engineering "all is heuristic," as Billy Vaughn Koen says. Until then you'll have to take my word for it.

FURTHER READING

Koen, B.V., 2003. Discussion of the Method. Oxford University Press, New York, NY.

APPENDIX 1

Integrated Design Example

I often use a particular Siemens ultrasonic level instrument to measure tank fluid levels. The instrument can display just the distance from ultrasonic transducer to liquid surface, or it can be programmed to display fluid level or even fluid volume (even with quite complex tank shapes).

It can output fluid level or volume as a 4–20 mA signal. It has five volt-free contacts which I can program to become active or inactive at various tank levels. It can have two transducers and be programmed to make decisions about which transducer's signal is the more reliable. It has all kinds of sophisticated signal processing onboard, and can be programmed as to which state to fail to if the signal from transducers is lost. It can produce alarm signals against many of these functions.

So if I want to control a set of duty/standby pumps in such a way as to maintain a constant tank level, and to avoid dry running, I have choices. The way I tended to use this instrument, back when it was less sophisticated, was to use it as a field-mounted standalone on/off level indicator controller. I used the volt-free contacts to start and stop pumps in succession, with the lowest set point providing dry running protection. The pumps were started under control of these contacts by online or star delta starters, and were either on or off.

I tend nowadays to use the instrument in one of two ways:

1. As a more sophisticated standalone modulating level indicator controller. I can wire the 4–20 mA output signal directly into a pump's inverter starter. Even motor starters have built-in computers nowadays, and an inverter drive can use the 4–20 mA signal to achieve all the same control actions as the last approach with no programmable logic controller (PLC) involvement.

2. As a dumb level indicator transmitter, with the 4–20 mA signal going out to the PLC, which then controls a number of pumps using variable speed drives (inverters) to go faster or slower in order to maintain a fairly constant tank level. I will probably under this scenario use one of the volt-free contacts as my hard-wired (direct into the pump starter) dry-running protection.

There are many other ways to use the instrument, but to take the three I have covered, the first and second have no PLC involvement, so they may save money on PLC costs, but with the drawback that they are less flexible as they cannot be so readily controlled via the PLC or system control and data acquisition (SCADA) system.

There is only one 4−20 mA output on the instrument, so if we wire it into the inverter, we cannot straightforwardly supply it to the PLC, so we cannot remotely monitor tank level, or intervene to alter action levels, and so on.

(A C + I specialist comments: "4−20 mA can be routed through more than one device as current loop; better practice is to use resistor to convert to voltage and parallel connect voltage signal into as many devices as you like. Inverters might have output available that could be configured as repeat. There are repeater isolator devices available.")

As I said, "cannot be so readily controlled."

So in this very common design scenario, which features at least once on virtually every plant I have designed, I tend to choose between these three options which I have used dozens of times before unless I can see a strong reason not to.

There is nothing wrong with starting the detailed design of process control systems by using standard control loops, strategies, and instruments, as it is close to how experienced professionals design. We have well-tried strategies in our heads, in much the same way as chess players do. Any inappropriate elements or unforeseen interactions between stock approaches should be picked up in design reviews and Hazard and Operability (HAZOP) studies.

At the design for construction stage, there are many interactions between the elements of design which experienced engineers will have anticipated at earlier design stages. In the example given, we are considering pumps and electromechanical components in addition to the software, controllers, and instruments which commonly fall under the heading of process control. We are informed more by broad experience than by a narrow but mathematically rigorous analysis of a few parameters.

INTEGRATED PROCESS CONTROL AND DESIGN EXAMPLE

To illustrate how we balance qualitative considerations, I will take an example from the water industry which I am very familiar with.

Consider the case of pretreatment of filter feed water. We need to add a coagulant to allow for the removal of colloidal matter, measure pH (and temperature to adjust for its effect on pH), and adjust pH to the minimum solubility of the product of our coagulant addition.

In theory this is very simple, but consider the following issues affecting process design/control:

Conceptual design issues

- The coagulant dose can be fixed, dosed proportional to flow, or dosed proportional to flow and color.

- Many coagulants are strongly acidic, so the coagulant dose will affect acid/alkali dose.
- We can choose lime or caustic for alkali addition. As lime has a greater buffering capacity, has an appreciable reaction time, and needs to be kept stirred if it is to be reasonably homogeneous, caustic is usually more controllable, but less forgiving of long loop times.
- We usually use hydrochloric or sulfuric acid for our acid addition. Adding chloride to the system may require a higher material specification.
- We usually need to control pH to within 0.1 pH units to give sufficient control over coagulant product solubility.
- We can dose chemicals using a number of pump types, or by gravity via a control valve.
- We need to have a short, sharp mix for both dosed coagulant and any required pH correction acid/alkali.
- We need to have a longer lower intensity mix for growth of the floc particles which the filter will remove.
- We can use static or dynamic mixing for either of these duties.
- Static mixers may not achieve the specified degree of mixing in the mixer body—the measurement point might need to be several pipe diameters downstream.
- Static mixers for chemical addition have a hydraulic residence time (HRT) measured in seconds. Dynamic mixers can have a HRT of several minutes.
- Both static and dynamic mixers for flocculation have a HRT of several minutes.
- There is no field-mounted instrument which reliably measures efficiency of flocculation, though we can measure it indirectly by measuring the turbidity which escapes the filter downstream, several minutes later.

Layout/piping issues

- We may need to have a certain length of straight pipe after the static mixer before the downstream pH sensor to ensure accuracy.
- We need to make sure the flow meters will run full, which may require their installation at a low point (avoiding dead legs), to ensure accuracy.
- The pH probe needs to be regularly removed and replaced for calibration, so it needs to be installed in such a way as to make this easy for operators.
- If we are going to use a gravity chemical addition system, we have to have a certain height available to us between storage tank and dose point.
- The output from the flocculator should be subjected to minimum shear prior to filtration to avoid breaking flocs.
- If we are going to use open-topped dynamic mixers, we will break head, so we need to arrange in the layout for them to flow to each other and then on to the filters by gravity, or (less favored) we need to add intermediate pumping stages.

Dosing issues

- Positive displacement pumps give a pulsating flow.
- Static mixers only mix radially.
- As a consequence of the last two issues, pulsation damping is strongly recommended if piston diaphragm (PD) pumps dose into a static mixer.
- Saturated sodium hydroxide solution freezes at 10°C.
- Precise control of chemical flow via a control valve on a line from a header tank is very hard to achieve, so we need to measure dosed flow if we are dosing via a control valve.
- In very dirty applications it might be difficult to keep the pH probes clean.
- Many pH probes now come with onboard temperature correction.
- Cheap and reliable field-mounted dedicated pH controllers are readily available.
- PD pumps are available which allow control of motor speed by one input 4−20 mA signal and control of stroke length via another.

Price issues

- Static mixers are far cheaper to buy and run than dynamic mixers.
- PD pumps are far more expensive to buy and run than solenoid pumps.
- Field-mounted pH controllers are a cheaper way to control pH than PLC control.
- Electromagnetic flow meters are more expensive to buy than turbine type, but are cheaper to run.

Safety issues

- Header tanks full of acids and alkalis at height carry safety concerns.
- The pressurized ring mains at height used to fill these header tanks carry safety concerns.
- We need to be sure that our main process pipework will not fill up with acid or alkali when the main plant shuts down, and that even if this did happen, this would not result in loss of containment.
- We need to be sure that positive displacement pumps are not throttled on suction or delivery sides, and that if making this impossible is unavoidable, over pressurization of delivery pipework does not lead to loss of containment.
- We need to be sure that even if our efforts to prevent loss of containment of acid/alkali fail, we have secondary containment in place.
- We need to be sure that if loss of containment occurs, and operators are contaminated, they can wash off the chemical ASAP.
- We need to be sure not to mix acid and alkali with each other.

Robustness issues

- If the operation of our plant depends upon effective coagulation, we will need standby units for all coagulant and acid/alkali dosing systems.
- PD pumps have a much better turn down, and are far more controllable than solenoid pumps.
- Peristaltic pumps are better than piston pumps for lime dosing, but their hoses present maintenance problems.
- Chemical addition is usually done via flexible lines. These lines have a limited life in service, and consideration needs to be given to their repeated replacement in operation.
- Systems based on dynamic mixing have far longer response times due to their much higher HRT.
- Electromagnetic flow meters are far more reliable and less prone to blockage than turbine type.
- Electromagnetic flowmeters are far more precise than turbine type.
- Flocculation can to some extent be optimized by measuring zeta potential, and particle size analyzers to measure filtration efficiency. However, in my opinion the kit to do this is, presently, more suited to lab than field (others disagree).
- We need to ensure dry running protection for any dosing pumps.
- We need to be able to tell if any powered mixers are operational.
- We need to know if any control valves are actually in the position we have requested.
- We need to know that the main process stream is flowing in order to prevent dosing into an empty line.
- Some flowmeters have empty pipe detection built in.

Integrated solution

So I would personally use a PD pump, with a 4–20 mA (or sometimes pulsed) signal from a field-mounted pH controller controlling stroke, and a 4–20 mA signal from an electromagnetic flowmeter in the main flow (with empty pipe detection) controlling speed for acid/alkali addition.

I would usually use caustic and sulfuric acid at around 40% w/w solutions for pH correction, though I sometimes use lower concentrations for operator safety reasons and to make freezing in the absence of tank heaters less likely.

The pump will deliver against a loading valve, and in between the loading valve and the pump there will be a suitably specified (especially in materials and capacity) pulsation damper vessel. Between the pump discharge and the first valve, there will be a return line to the feed tank protected by a pressure relief valve (PRV). This arrangement shall be so arranged that if the PRV lifts, the return flow will be visible to operators.

The pH signal in to the controller comes from a probe, mounted five pipe diameters of unobstructed pipe downstream of a static mixer, specified to give CoV (Co-efficient of Variance) = 0.05 (or less technically a 95% degree of mixing) at that measurement point across the operating flow range.

The static mixer has separate connections for coagulant, acid, and alkali. The chemicals are dosed through injection lances (incorporating spring-loaded nonreturn valves) into the centerline of the pipe. Hoses should be contained in trace-heated and lagged pipes which provide a duct to pull though a replacement hose, as well as secondary containment.

Coagulant dose is via a PD pump (or a solenoid pump if the plant is small and price is crucial), controlled by a 4−20 mA signal direct from an electromagnetic flowmeter.

Chemicals will be stored in quantities suitable for at least a month, with external connections suited to suppliers working in that area. In the United Kingdom we might need a tank heater, and will probably need trace heating and lagging of NaOH systems at a minimum. Ideally all lines which might freeze under all reasonably foreseeable conditions will be trace heated and lagged. (I have had clients talk me out of this requirement in the past, and they have had cause in time to regret the penny-pinching.)

The dosed main process flow will pass from a static mixer to a static flocculator, selected for robustness and simplicity. Line sizes and fittings from flocculator to filter shall be selected to minimize headloss and therefore shear.

(I have been told by a correspondent that digital dosing pumps with integrated speed/stroke control have replaced simple PD pumps—it only takes a few years for the state of the art to move on).

So Luyben's idea of incorporating controllability into design is sound, but there is a lot more to controllability than math or generic theory. The insights of plant operators and commissioning engineers and their knowledge of the detailed characteristics of kit need to be incorporated into the design.

Some might think that some of the things in the description I give above are not control issues, but they encapsulate choices as to whether to solve a design problem with soft or hard control. For example, if I do not ensure that my pipes will not freeze through physical arrangement, soft controls will be needed.

APPENDIX 2

Upset Conditions Table

This table, and the associated text, was first published in "Process Plant Design and Operation," Doug Scott & Frank Crawley, pp. 94–105, Copyright Elsevier 1992.

SPECIFIC UPSET CONDITIONS

In the pages that follow, upset conditions are identified by the guidewords used in hazard and operability studies—this is where upsets are most likely to be identified prior to plant operation. The upset condition can then be analyzed under the four phases of the project: conceptual (Con); detailed design (Det); startup (St), and operation (Op). For ease of presentation and analysis, the conditions are analyzed in block format with possible solutions and individual reference numbers. The time when the upset condition is most likely to be detected is indicated by an "X" in the project phase. In this manner, the upset condition can be read vertically in the matrix and the project timing can be read horizontally.

It should be noted that the proposed solution may not always be applicable to a specific problem, but they all have been used at one time or another. Any tabulation of this type can only give guidance. It can never hope to be comprehensive. In some instances conflicting requirements may be identified and engineering judgment will be necessary to ensure the most appropriate solution is chosen.

Table A2.1 Specific upset conditions
Pressure

No	Condition	Solution	Most likely stage			
			Con	Det	St	Op
P1	More pressure	Devise a process which operates at or near atmospheric pressure and does not utilize volatile fluids or vigorous reactions.	X			
P2	More pressure	Specify the design pressure of equipment such that it cannot be overpressured by any condition other than fire.	X	X		
P3	More pressure	Consider the potential for metal fatigue following pressure cycling.		X		X

(Continued)

Table A2.1 (Continued)
Pressure

No	Condition	Solution	Most likely stage			
			Con	Det	St	Op
P4	More pressure	Take due account of elevation changes and design for the sum of both hydraulic head and vapor pressure (see L1).	X	X	X	
P5	More pressure	Install a high integrity protective system to shut the process down before the overpressure condition is encountered.		X		X
P6	More pressure	Install a full rated safety relief system and analyses the means by which the fluids may be dispersed in they are toxic or flammable.		X		
P7	More pressure	Steam trace or, purge relief valve nozzles to prevent the deposition of foulants.		X	X	X
P8	More pressure	Specify the failure action of control systems so as to minimize the effect of failure (see P16).		X	X	X
P9	More pressure	Initiate control procedures such that flow limiting devices such as orifice plates and control valves can only be changed after a safety study has been carried out (see F4 and OP2).		X	X	
P10	More pressure	Carry out routine proof tests of relief valves and test facilities for protective systems.		X	X	X
P11	More pressure	Install duplicate relief valves and test facilities for protective systems.		X		X
P12	More pressure	Rod through vents to ensure that lines are clear of debris. Check flame arrestors are clear and not choked with debris (see F8, F9 and P21).		X	X	X
P13	More pressure	Install purge points on total condensers to allow the removal of inert gases like air and nitrogen.		X	X	X
P14	No pressure	Choose a process which will not reach a hazardous condition if pressure falls or is lost; this may apply to oxidation processes where oxygen and hydrocarbons could enter the flammable regime should the reaction stop.	X			

(Continued)

Table A2.1 (Continued)
Pressure

No	Condition	Solution	Most likely stage			
			Con	Det	St	Op
P15	No pressure	Specify the metallurgy such that the metal will not enter a brittle regime when depressured. This might apply to cryogenics and refrigerants (see T13).		X		
P16	No pressure	Specify failure action of control systems so as to minimize the effect of the failure. This may be contrary to the needs of more pressure (see P8).		X	X	X
P17	Less pressure	Consider the effects of leaks in vacuum condensers. This is a variant of no pressure.		X	X	X
P18	Reverse pressure	Design equipment for full pressure and vacuum, including liquid head.		X		
P19	Reverse pressure	Specify pressures within the process such that leakage across heat exchangers produces a safe condition, e.g. steam leaks into hydrocarbons and not hydrocarbon leaks into steam.		X		X
P20	Reverse pressure	Install vacuum protection where appropriate e.g. on fixed roof storage tanks.		X		
P21	Reverse pressure	Rod out vents to prove they are clear. Check flame arrestors for debris (see P12, F8 and F9).		X	X	X
P22	Reverse pressure	Check vacuum relief valves for operation.			X	X
P23	Reverse pressure	Be mindful of the causes of vacuum: 1. Sucking in suction catch pots when air testing compressors		X	X	X
		2. Draining vessels			X	X
		3. Steaming out vessels				X
		4. Adding cold fluids to hot vessels (rapid condensation)			X	X
		5. Internal reactions causing volume shrinkage (e.g. polymerization or rusting)			X	X
		6. Polythene sheets blowing over vents and breather lines.			X	X

(Continued)

Table A2.1 (Continued)
Level

No	Condition	Solution	Most likely stage			
			Con	Det	St	Op
L1	More level	Specify the design temperature of equipment for the sum of hydraulic head and vapor pressure (see P4).		X		
L2	More level	Locate vapor relief valves at the top of the equipment so that they are not 'drowned' by liquid.		X		
L3	More level	Consider stressing pipelines and pipe supports for the liquid full condition (required for hydraulic pressure testing).		X		
L4	More level	Consider the loading on foundations and structures during hydrotesting. If the vessel is totally flooded, design for the worst case.		X		
L5	More level	Changes in interface level may result in separate liquid phases passing forward along the process route, should protective systems be installed? (see L7 and OT10).		X	X	X
L6	More level	If equipment sizes are increased for any reason consider the extra loading on supports, particularly during hydrotesting.			X	X
L7	Less level	The light phase will pass forward as entrained fluid (see L5 and OT10).		X	X	X
L8	Less level	Electric heaters or temperature probes may be exposed. Low level trips should be fitted to cut off the power (see L11 and T5).		X	X	X
L9	No level	Will the loss of level result in the loss of liquid flow to a vital system such as cooling water, lubricating oil or seal oil? Should a protective system be installed? (See OT10).		X	X	X
L10	No level	Will the loss of level result in a gas 'blow by' from a high to a low pressure system? Consider installing flow chokes and protective system or full flow pressure relief on the low pressure system.		X		X

(Continued)

Table A2.1 (Continued)
Level

No	Condition	Solution	Most likely stage			
			Con	Det	St	Op
L11	No level	Will the loss of level result in overheating? Consider the effect of loss of level in a boiler, a reboiler or an electrically heated vessel. Install low level trips (see L8 and T5).		X	X	X
L12	No level	Install bunds round storage tanks sized for 1.1 times the storage tank capacity for containment in the event of tank rupture.		X		
L13	Reverse level	Consider splash filling vessels at a higher elevation as opposed to filling under liquid levels and possibly causing a syphon effect. However, consider the generation of static electricity if the fluids are flammable.		X	X	X
L14	Reverse level	Consider reverse level (i.e. from high to low level) as a potential for reverse flow.		X	X	X

Temperature

No	Condition	Solution	Most likely stage			
			Con	Det	St	Op
T1	More temperature	Is the reaction exothermic? Can the reactor 'run away'? Consider the need for protective systems such as quenching, catalyst kill or the equivalent.	X	X		
T2	More temperature	Size the reactor cooling system with excess capacity to prevent a run away. Ensure that mixers have a reliable power supply.		X		
T3	More temperature	Consider the potential for metal fatigue due to temperature cycling.		X		X
T4	More temperature	1. Install low flow trips in fired heaters		X		X
		2. Install high stack temperature alarms in fired heaters		X		X
		3. Install high metal temperature alarms in fired or electrically powered heaters.		X		X

(*Continued*)

Table A2.1 (Continued)
Temperature

No	Condition	Solution	Most likely stage			
			Con	Det	St	Op
T5	More temperature	1. Install low flow trips in electrically heated systems		X		X
		2. Install low level trips in electrically heated vessels (see L8 and L11).		X		X
T6	More temperature	Specify materials of construction to give adequate allowance for creep.		X		X
T7	More temperature	Inspect equipment for evidence of creep on a regular basis. Note, evidence of creep may manifest itself suddenly after a number of years of operation. Creep is a cumulative effect — a number of short deviations may lead to serious damage in the future.				X
T8	More temperature	Specify failure action of control valves so as to minimize the effect of failure.		X	X	X
T9	More temperature	Consider the potential for overheating when pumps or compressors are blocked in.		X	X	X
T10	Less temperature	Consider the effects of freezing in cold environments (see C4). Water can freeze in drain lines, instrument trappings, relief valves, valve bonnets, pump casings, low point and fire water lines. Examine the need for heat tracing, thermal insulation draining, maintaining a small flow of fluids to maintain a limited heat input.		X	X	X
T11	Less temperature	Consider the possible effects of crystallization of process fluids in relief operations and emergency drain/blow down systems. Should these be heat traced? (See C2)		X	X	X

(Continued)

Table A2.1 (Continued)
Temperature

No	Condition	Solution	Most likely stage			
			Con	Det	St	Op
T12	Less temperature	Consider the possible effects of gels of high viscosity. Should lines be heat traced? (see C3)		X	X	X
T13	Less temperature	Specify the metallurgy such that the metals will not enter a brittle regime when depressured (see P15).		X		X
T14	Less temperature	Specify the failure action of the control valve so as to minimize the effect of failure.		X		X

Flow

No	Condition	Solution	Most likely stage			
			Con	Det	St	Op
F1	More flow	Consider the effect of rise or fall of levels in equipment.		X	X	X
F2	More flow	Consider the potential for exciting tube vibration in heat exchangers and tube failure caused by fatigue or wear from high velocity.		X		X
F3	More flow	Consider the potential for exciting vibration in thermowells.		X		X
F4	More flow	1. Install flow limiting devices		X	X	X
		2. Size relief systems for full flow through the flow limiting device		X	X	X
		3. Register the flow limiting device as a protective system (see P9 and OP2).			X	X
F5	More flow	Consider the potential for erosion in bends due to solids or droplets of liquids in gases.		X		X
F6	More flow	Install shallow bunded areas round pumps, fired heaters and heat exchangers to cater for spillage and to retain foam blankets in fires.		X		X
F7	No flow	Install low flow trips on fired heaters or electrically heated systems.		X		X
F8	No flow	Monitor flame arrestors for fouling (see P12 and P21).			X	X

(*Continued*)

Table A2.1 (Continued)
Flow

No	Condition	Solution	Most likely stage			
			Con	Det	St	Op
F9	No flow	Rod vents to prove they are clear (see P12 and P21).			X	X
F10	No flow	Do not install temperature measurement points in areas of no flow.		X		
F11	Reverse flow	Install non-return valves in pumped systems, in potential siphons, in flexible loading/off-loading systems.		X	X	X
F12	Reverse flow	Can fluids be passed from one section of the plant to another via drain or vent/blow down systems? Consider the potential hazards from flow and mixing of incompatible fluids.		X	X	X
F13	Reverse flow	If a drain or vent system is choked, can fluids pass from a high to a low pressure vessel?		X	X	X
F14	Reverse flow	Check the size of vent headers to ensure that the pressure drop down the header does not result in overpressure or low pressure equipment.		X		X
F15	Reverse flow	Can air be drawn into a hydrocarbon system due to condensation, process upset or flow regimes?		X	X	X

Concentration

No	Condition	Solution	Most likely stage			
			Con	Det	St	Op
C1	More concentration	Consider what may happen if the concentration of any reactant or catalyst arises. Will the reactor produce unwanted by-products or become unstable? What warning is needed?		X	X	X
C2	More concentration	Can solids crystallize out of a liquid phase? (see T11)		X		X
C3	More concentration	Can fluids become waxy or very viscous? (see T12)		X		X

(Continued)

Table A2.1 (Continued)
Concentration

No	Condition	Solution	Most likely stage			
			Con	Det	St	Op
C4	More concentration	Consider the effects of deposits on instrument tappings, relief systems and drain systems (see T10).		X	X	X
C5	More concentration	Consider the effects of higher or lower pH on metallurgy. It may be prudent to assume that higher concentrates may occur.		X	X	X
C6	More concentration	Consider the possibility of build-up in the concentration of impurities in reactors, reboilers and condensers. Should purge systems be installed?		X	X	X
C7	More concentration	Consider the possibility of erosion in slurry systems. Should bends be installed with extra wall thickness? Should flushing points be fitted?		X		X
C8	More concentration	Consider the possible detrimental effects of concentrated aqueous spills or leakages into thermal insulation. Could there be acid/alkali/salt concentration which will attack metal and cause stress corrosion cracking?		X		X
C9	More concentration	Consider the possibility of concentration of toxics or unstable chemicals in the process, e.g. acetylenes are particularly unstable in high concentrates.		X		X
C10	More concentration	Consider the adverse reactions that may take place if heat exchangers leak. Could this affect the process, the reactor chemistry or the metallurgy?		X		X
C11	More concentration	Do control variables such as reflux or reboil require resetting if concentrations change?		X	X	X
C12	More concentration	What are the maximum ground level concentrations from vents? Are they safe? Are they unpleasant/offensive? Should the vents lead to a flare for pyrolysis? Should the stack height be increased?		X		X

(Continued)

Table A2.1 (Continued)
Concentration

No	Condition	Solution	Most likely stage			
			Con	Det	St	Op
C13	More concentration	If the effluent concentration changes can it create environmental pollution?		X	X	X
C14	Less concentration	Will the reaction stop if the concentration of reactants or catalyst falls? What warning is needed?		X		X
C15	Less concentration	Do dilute concentrations create excessive heat loads in separation systems?		X	X	X
C16	More and less concentration	Can the process enter a flammable regime during normal or upset operation? Consider startup/ shutdown and condensation.	X	X	X	X

Other than

This is a variable which requires the most careful consideration as it has so many disguises. Some have already been addressed in other sections, but it should not be assumed that all have been identified.

No	Condition/Solution	Most likely stage			
		Con	Det	St	Op
OT1	Creep (see T7)				X
OT2	Corrosion (see C5)			X	X
OT3	Erosion (see F5)		X		X
OT4	How will equipment age: do joints soften, harden or crack? Will joints fail prematurely?		X		X
OT5	Are minor components of the process incompatible with process fluids, such as copper gaskets in ammonia?		X		X
OT6	Can fluids generate static charges under flow conditions?		X	X	X
OT7	What by-products may be expected, e.g.	X	X	X	X
	Pyrites	X	X	X	X
	Polymers: unstable, explosive	X	X	X	X
	Polymers: unstable, hydrolysis to toxic gases	X	X	X	X

(Continued)

(Continued)

No	Condition/Solution	Con	Det	St	Op
			Most likely stage		
OT8	Are contaminants (air and water) positive catalysts or inhibitors?		X	X	X
OT9	Is there a risk of hydrogen blistering or other unexpected corrosion effects?		X		X
OT10	What happens if levels of liquid/liquid interface levels are lost? (see L5, L7 and L9)	X	X	X	X
OT11	Is air a potential 'other than' in the presence of flammables?		X	X	X

Other than − failure

No	Condition/solution	Con	Det	St	Op
			Most likely stage		
FA1	How does equipment fail?				
	1. Pump seals leak − is this tolerable?		X	X	X
	2. Heat exchangers corrode, erode, wear and fatigue		X		X
	3. Vessels corrode and pit				X
	4. Structures rust (and corrode in acid environments even under lagging)	X	X		X
	5. Bearings or rotating equipment collapse − will this create an intolerable seal leak?		X	X	X
FA2	What is the effect of instrument air failure, both local or plant wide? Will the plant shut down safely?		X	X	X
FA3	What is the effect of service failure such as cooling water or nitrogen purge?		X	X	X

Operations

No	Condition/solution	Con	Det	St	Op
			Most likely stage		
OP1	What controls are imposed to prevent staff using unsafe operational practices rather than operating procedures e.g. audits, site tours, casual enquiries?			X	X
OP2	What controls are imposed to prevent changes in design intent? Are modification control procedures in place? (see F4 and P9)			X	X
OP3	What controls are in place to prevent the override of trips/protective systems? Is the trip test system 'operator resistant'?		X	X	X
OP4	What controls are in place to ensure that protective systems e.g. relief valves and trips, are tested routinely?			X	X

(Continued)

(Continued)

Operations

No	Condition/solution	Most likely stage			
		Con	Det	St	Op
OP5	What controls are in place to ensure that flow limiting devices are not removed?			X	X
OP6	Are operating instructions rewritten on a routine basis? Are all operators aware of changes?			X	X
OP7	Are maintenance procedures written and followed correctly?			X	X
OP8	Is there adequate communication between control centers and outside operators?		X	X	X
OP9	Are audit procedures in place?			X	X

APPENDIX 3

Plant Separation Tables

Text and accompanying tables reproduced and adapted from "Process Plant Layout," edited by J.C. Mecklenburgh, Copyright IChemE 1985.

Detailed design requirements are to be found in the appropriate national codes, standards, and specifications for the individual classes of equipment and operation. The following tables provide some guidelines of typical restraints which might be applied during the early development of the site and plant layout with due consideration given to safety.

Some of these typical quantities are taken from old codes of practice, etc. which have now been superseded or withdrawn because the uncritical use of standard distances is no longer accepted as good layout practice. However, these standard distances in the old publications can still be considered as good typical values for initial layout.

It must be emphasized, though, that no values in this appendix should appear in the final layout without detailed checking that they apply to the circumstances of the plant or site being designed.

During the final stages of preparing this book, I have been engaged to update Mecklenburgh. I will revise the tables in my updated version to meet modern regulatory requirements. Until then, the following data is the best set in the public domain, and is still a sound starting point.

Table A3.1 Site areas and sizes (preliminary)

Administration	10 m^2	Per administration employee
Workshop	20 m^2	Per workshop employee
Laboratory	20 m^2	Per laboratory employee
Canteen	1 m^2	Per dining space
	3.5 m^2	Per place including kitchen and store
Medical Centre	0.1–0.15 m^2, minimum 10 m^2	Per employee depending on complexity of service
Fire Station (housing 1 fire, 1 crash, 1 foam, 1 generator and 1 security vehicle)	500 m^2	Per site
Garage (including maintenance)	100 m^2	Per vehicle
Main perimeter roads	10 m	Wide
Primary access roads	6 m	Wide
Secondary access roads	3.5 m	Wide
Pump access roads	3.0 m	Wide
Pathways	1.2 m	Wide up to 10 people/min.
	2.0 m	Wide over 10 people/min. (e.g. near offices, canteens, bus stops)
Stairways	1.0 m	Wide including stringers
Landings (in direction of stairway)	1.0 m	Wide including stringers
Platforms	1.0 m	Wide including stringers
Road turning circles – 90 degree turn and T-junctions		Radius equal to width of road
Minimum railway curve	56 m	Inside curve radius
Cooling towers per tower	0.04 m^2/kW to 0.08 m^2/kW	Mechanical draught Natural draught
Boiler (excluding house)	0.002 m^3/kW	(height = 4 × side)

Table A3.2 Preliminary general spacings for plots and sites

Item	Property boundary	Control room (non-pressurized)	Control room (pressurized)	Administration building	Main Substation	Shippings, buildings, warehouses	Loading facilities, road, rail, water	Fire pumphouse	Cooling Towers	Process Fired heaters	Gas Compressors	Reactors	High pressure storage spheres, bullets	Atmospheric flammable liquid storage tanks	Air coolers	Low pressure storage spheres or tanks < 1 bar G	Plot limits	Process control station	Process unit substation	Process equipment (low flash point)	Process equipment (high flash point)
Cryogenic O$_2$ plant^a	30																				
Process equipment (high flash point)	2	CP																			
Process equipment (low flash point)	2	2	CP																		
Process unit substation	NA	15	15	50																	
Process control station	NA	NM	15	15	60																
Plot limits	15	NA	NM	NA	NA	CP															
Low pressure storage spheres or tanks < 1 bar G	CP	CP	60	NM	CP	CP	CP														
Air coolers	NM	60	NA	15	15	5	5	30													
Atmospheric flammable liquid storage tanks	CP	60	CP	CP	60	NM	CP	CP	CP												
High pressure storage spheres, bullets	CP	CP	60	CP	CP	60	NM	CP	CP												
Reactors	2	60	5	60	NA	15	15	5	5	60											
Gas Compressors	2	10	75	60	7.5	60	NA	15	15	7.5	7.5	45									
Process Fired heaters	7.5	15	15	CP	15	CP	NA	15	15	30	15	CP									
Cooling Towers	7.5	30	30	30	30	30	30	30	15	15	30	30	30								
Fire pumphouse	NA	30	60	60	60	75	60	60	60	NA	NM	60	60	45							
Loading facilities, road, rail, water	NA	45	45	60	60	60	CP	CP	60	CP	45	60	45	60	CP						
Shippings, buildings, warehouses	NA	60	30	30	75	60	60	75	60	60	60	NA	NA	60	60	60					
Main Substation	NA	15	60	30	60	60	60	75	60	60	60	60	30	NM	60	60	60				
Administration building	NA	8	30	60	30	75	60	60	75	60	60	75	60	NA	NA	60	60	30			
Control room (pressurized)	NA	8	15	NM	30	8	30	30	30	30	30	30	30	60	NA	NA	NM	NM	NM		
Control room (non-pressurized)	NA	NA	8	30	15	60	8	30	30	30	30	60	60	30	60	NA	NM	30	30	30	
Property boundary	NA	30	8	8	8	30	8	30	30	30	60	CP	CP	60	CP	60	30	NM	15	15	CP

Legend:
NA Not applicable since no measurable distance can be determined.
NM No minimum spacing established – use engineering judgment.
CP Reference must be made to relevant Codes of Practice but see the following section on Preliminary Spacings for Tank Farm Layout.
^{a}Also see the following section on Preliminary Spacings for Tank Farm Layout for minimum clearances.

Notes:

(a) Flare spacing should be based on heat intensity with a minimum space of 60 m from equipment containing hydrocarbons

(b) The minimum spacings can be down to one-quarter these typical spacings when properly assessed.

Table A3.3 Preliminary access requirements at equipment

Access	Item of equipment
Permanent Ladder	1. Gate and globe valves — DN 80 (in mm) and smaller at vessels when located 3.5 m above grade 2. Check valves — all sizes at vessels when located 3.5 m above grade 3. Gauge glass 2 m above access surface, or inaccessible by portable ladder or platforms 4. Pressure instrument on vessels 2 m above access surface, or inaccessible by portable ladder or platform 5. Temperature instrument on vessels 2 m above access surface, or inaccessible by portable ladder or platform 6. Handholds located 3.5 m above grade

Items located over platform

Platforms	7. Manholes 8. Heat Exchange Units 9. Process blinds 10. Relief valves on vertical vessels DN100 and larger 11. Control valves — all sizes 12. Cleanout points

Items located adjacent to platform

	13. Gate and globe valves — DN100 and larger at vessels 14. Motor operated valves 15. Relief valves — DN80 and smaller 16. Relief valves on horizontal vessels DN 100 and larger 17. Level controls an gauge glass on vessels 18. Sampling valves on vessels

Table A3.4 Preliminary minimum clearances at equipment

Item		Description	Clearance (m)
Roads	1.	Headroom for primary access roads or major maintenance vehicles	6.0
	2.	Width of primary access roads	6.0
	3.	Headroom for secondary roads and pump access roads	3.0–4.5
	4.	Width of secondary roads and pump access roads	3.0–4.5
Railways	5.	Headroom over through railways from top of rail	6.7
	6.	Headroom over dead-ends and sidings from top of rail	5.1
	7.	Clearance from track centerline to obstructions	2.4
Access, walkways and maintenance clearances	8.	Headroom over platforms, walkways, access ways, maintenance areas	2.5
	9.	Width of stairways, back to back of stringers	0.75
	10.	Width of landings in direction of stairway	0.9
	11.	Width of walkways at grade or elevated	0.75
	12.	Vertical rise of stairways – one flight	4.5
	13.	Vertical rise of ladders – single run	7.5
	14.	Clearance under furnace burner nozzles for maintenance purposes	2.1
Platforms	15. Towers,	Distance of platform below bottom of manhole flange (side platform)	0.3
	16. Vertical	Width of manhole platforms from manhole cover to outside edge of platform	0.75
	17. and	Platform extension beyond centerline of manhole flange (side platform)	0.75
	18. horizontal	Distance of platform below underside of flange (top platform)	0.2
	19. vessels	Width of platform from three sides of the manhole (top platform)	0.75
	20. Horizontal	Clearance in front of channel or bonnet flange	1.2
	21. exchangers	Clearance from edge of flanges	0.3
	22. Vertical	Distance of platform below top flange of channel or bonnet	1.5 max
	23. exchanger	Width of platform from three sides of flange	0.6
	24. Furnaces	Width of platform at sides of horizontal and vertical tube furnace	0.75
	25.	Width of platform at ends of horizontal tube furnaces	1.0
Pipeways	26.	Pipeways not crossing roads	3.0

Table A3.5 Handling facilities for equipment

Item		Equipment and equipment part handled	Handling facility
Vertical vessels	1.	Manhole covers (up to DN 600) and vessel trays	Davits
	2.	Bottom manholes	Hinged
	3.	Internals of fixed bed reactors, catalyst, tower packings etc.	None
Horizontal exchangers (at grade or in structure)	4.	Removable tube bundles, and other removable parts except exchanger shells, shell covers, and floating head covers	Pulling beams or posts, for moving the bundle within the shell. Trolley beams for groups requiring up to four such beams. Trolley beams shall be provided with either (a) two trolleys, one capable of handling the entire load and the other half-capacity, or (b two half-capacity trolleys
	5.	Exchanger shells	None
	6.	Fixed tube sheet exchangers	None
		Shell covers and floating head covers	Shell davits or overhead hitching points
Vertical exchangers	7.	Stationary tube sheet at lower end — Tube bundles, channels, and channel covers	Hitching points
	8.	Shell covers and floating head covers	Jib crane, davit, or hitching point
	9.	Entire small-size units	Hitching point or trolley beam
	10.	Stationery tube sheet at upper end — Units designed for removing tube bundle from shell — Tube bundles, channels and channel covers	Trolley beam
	11.	Shell covers and floating head covers	Hitching points
	12.	Entire small-sized units	Hitching point
	13.	Units designed for removing the shell from the bundle: the entire unit or any of its component parts	Hitching point
	14.	Fixed tube sheet exchangers	Shell davits or hitching points

Pumps, compressors and drivers (housed or otherwise inaccessible)	15.	100 kg—2t incl.	Parts of horizontal centrifugal pumps and steam drivers	Overhead hitching point or trolley beam
	16.		Cylinder heads and pistons only of reciprocating pumps and horizontal reciprocating compressors	
	17.	Over 2t	Parts of centrifugal pumps, compressors and steam drivers including top halves of compressors, and turbine covers	Trolley beam or overhead travelling cable
	18.		Cylinder heads and pistons only of reciprocating compressors	
	19.		Power cylinders only of inclined type reciprocating compressors	
Piping (housed or otherwise inaccessible)	20.		Parts of vertical-type pumps and drivers	Overhead hitching point
	21.		Electric motors and rotors	None
	22.		Relief valves, DN 100 × 150 and larger	Hitching points or davits
	23.		Blanks, blind flanges, fittings, and valves other than listed above and weighing more than 150 kg	Hitching points or davits when subject to frequent removal for operation or maintenance

PRELIMINARY SPACINGS FOR TANK FARM LAYOUT

1. Where space allows, greater distances than these should be used. The incorporation of these minimum distances into a design can only be made after a proper assessment.
2. Flammable liquids for this table are defined as those with flash points up to 66°c.
3. Measured in plan from the nearest point of the vessel, or from associated fittings from which an escape can occur when these are located away from the vessel.
4. A group of vessels should not exceed 10000 m^3 unless a single vessel. Spacing between such groups should be a minimum of 15 m between adjacent vessels. The bund to have a net volume not less than 10% of the capacity of the largest tank in the bund after deducting volume up to bund height of all other tanks in the same bund.
5. If this distance cannot be achieved, the need for suitable fire protection of the cable or pipeline should be considered.
6. For bunded tanks containing water-soluble non-hydrocarbons, power cables and pipelines at ground level should be outside the bund and so protected by the fund from fire in the tanks.
7. Measured in plant from the nearest part of the bund wall except where otherwise indicated.
8. A group of tanks should not exceed 60000 m^3. Spacing of the nearest tanks in any two such groups, which may have a common bund wall, should be such that the tank in one group should be a minimum of 15 m from the inside top of the bund of any adjacent group(s).
9. The zone may be beveled across its upper corners providing all parts of the vessel more than 3 m from the zone edge.

Table A3.6 Preliminary minimum distances (note 1) for liquefied oxygen (5,6)

	Distance (m)
To site boundary	30
To site roads	15
To process units and buildings containing combustible materials and ignition sources	30
To outside fixed combustible materials	5
To buildings containing flammable fluids	45
To road and rail loading areas	15
To overhead power lines and pipebridges	30
To other above-ground cables and important pipelines or pipelines containing flammables	15
To underground cables, trenches	10
To low-pressure gas storage	30
To compressed gas storage: flammable	30
non-flammable	15
To liquefied pressure and refrigerated storage: flammable	45
non-flammable	15
To liquid storage tanks: flammable (note 2)	45
non-flammable (note 2)	30

Table A3.7 Preliminary minimum distances (note 1) for liquefied, flammable gases

Item	Material stored		
	Hydrocarbons	*Non-hydrocarbons insoluble in water*	*Non-hydrocarbons soluble in water*
Pressure storage (Notes 3,4)			
To boundary, process units, buildings containing a source of ignition, or any other fixed sources of ignition, e.g. process heaters	For example: Ethylene 60 m C_3 45 m C_4 30 m	For example: Methyl Chloride 23 m Vinyl Chloride 23 m Methyl-vinyl ether 23 m Ethyl Chloride 15 m	For example: Methylamines 15 m
To building containing flammable materials e.g. filling shed	15 m	15 m	15 m
To road or rail tank wagon filling points	15 m	15 m	15 m
To overhead power lines and pipebridges	15 m	15 m	15 m
To other above-ground power cables and important pipelines or pipelines likely to increase the hazard	(Note 5) 7.5 m	(Note 5) 7.5 m	See Note 6
Between pressure storage vessels	One-quarter of sum of diameters of adjacent tanks but not less than 1.8 m for ≤ 50 m^3 or less than 15 m for 750 m^3		
To low pressure refrigerated tanks	15 m from the bund wall of the low pressure tank, but not less than 30 m rom the low pressure tank shell		
To flammable liquid (note 2) storage tanks	15 m from the bund wall of the flammable liquid tank		
To liquid oxygen storage	As defined above under 'Liquefied Oxygen'		
Zone 1 extent	1 m sphere around relief valve discharge		
Zone 2 horizontal extent from edge of tank	For example: Ethylene 30 m C_3's 30 m C_4's 20 m	For example: Methyl Chloride 15 m Vinyl Chloride 15 m Methyl-vinyl ether 15 m Ethyl Chloride 10 m	For example: Methylamines 10 m
Zone 2 height of zone	260 × relief diameter above relief valve discharge (see note 9)		

(Continued)

Table A3.7 (Continued)

Item	Material stored		
	Hydrocarbons	Non-hydrocarbons insoluble in water	Non-hydrocarbons soluble in water
Low pressure refrigerated storage (Notes 7,8)			
To boundary, process units, buildings containing a source of ignition, or any other fixed sources of ignition	For example: Ethylene 90 m C$_3$'s 45 m C$_4$'s 15 m		For example: Ethylene oxide 15 m
To building containing flammable materials e.g. filling shed	15 m		15 m
To road or rail tanker filling point	15 m		15 m
To overhead power lines and pipebridges	15 m		15 m
Between low pressure refrigerated tanks	One-half of sum of diameters of adjacent tanks		
To flammable liquid (note 2) storage tanks	Not less than 30 m between low pressure refrigerated LFG and flammable liquid tank shells, but LFG and flammable liquids must be in separate bunds		
To pressure storage vessels	As defined above under 'Pressure Storage'		
To liquid oxygen storage	As defined above under 'Liquefied Oxygen'		
Zones 1 and 2	As defined above under 'Pressure Storage'		

Table A3.8 Liquids stored at ambient temperature and pressure

Preliminary minimum clearance

Dim (Fig A3.1 & A3.2)	Diameter of tank	Water and non-flammable liquids	Class A and B Products — Fixed roof	Class A and B products (flash point <32°c) — Floating roof	Class C products
A	Up to 6 m	—	3 m		3 m
	6–30 m		Half tank dia.	6 m	Half tank dia.
	Over 30 m		15 m		Half tank dia.
B	All	1.5 m	Least of: Half dia. of largest tank, dia. of smallest tank, 15 m. (min 6 m)	Least of: Half dia. of largest tank, 6 m	Half tank dia. Smallest tank Min 3 m
C	All	—	Dia. of largest tank Min 10 m	Dia. of largest tank (Min 6 m)	Dia. Of largest tank Min 6 m
D	All	6 m	15 m	6 m	6 m
E	All	—	15 m	6 m	6 m
F	All	6 m	7.5 m	7.5 m	7.5 m
G	All	5 m	30 m	30 m	15 m
H	All	Depends on building lines	30 m	30 m	15 m
J	All	—	30 m	30 m	15 m
K	All	—	15 m	15 m	15 m
M	All	—	15 m	15 m	15 m
N	All		7.5 m	7.5 m	7.5 m
P	All	Outside bund	Outside bund	Outside bund	Outside bund
Q	All		Bund Width	Bund Width	Bund Width
U	Up to 3.5 m		1.5 m	3 m	—
	Over 3.5 m		3 m	3 m	—
V	All		3 m	3 m	—
W	All		All bund to wall height	All bund to wall height	—
X	Up to 3.5 m		5 m	15 m	—
	3.5–5 m		6 m		—
	Over 5 m		15 m		—
Y	Up to 3.5 m		2 m		—
	3.5–5 m		2.5 m		—
	Over 5 m		5 m	5 m	—
Z	All		5 m	5 m	—
Max capacity/bund			60000 m³	120000 m³	—

The spacing and arrangement of tankage can vary with each application (note 1).
Class A products have closed flash points below 23°c.
Class B products have closed flash points between 23–66°c.
Class c products have closed flash points above 66°c.

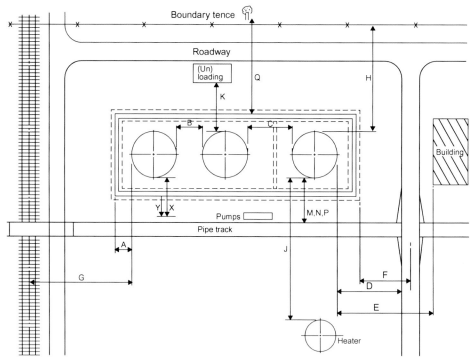

Figure A3.1 Preliminary tank farm layout (A) plan view.

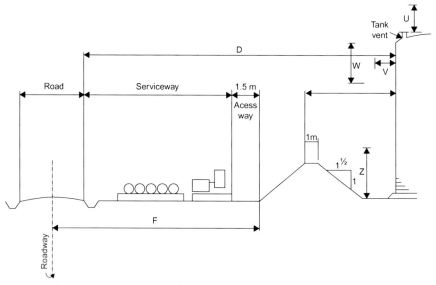

Figure A3.2 Preliminary tank farm layout (B) elevation.

Legend	Distance
A	from outside of tank to outside of bund at top
B	between any 2 tanks in one tank bund
C	between any two tanks in adjacent bunds
D	from tanks to main plant roads
E	between tanks and buildings containing flammable material
F	from toe of bund to center line of any plant roads
G	from tank to center of railway
H	from tank to boundary fence
J	between tank and fired heaters or ignition sources
K	from tank to road or rail filling
M	from tank to ground underneath power lines and pipe bridges
N	from tank to power cables or pipelines
P	from tank to ground above buried cables or pipes
Q	from tank to combustible materials
U	from tank vent to top of zone 1
V	from outside of tank to edge of zone 1 and zone 2
W	from tank rim to junction of zone 1 and zone 2
X	from outside of tank to edge of zone 2
Y	from center line of bund wall to edge of zone 2
Z	from ground to top of zone 2

PRELIMINARY ELECTRICAL AREA CLASSIFICATION DISTANCES

Note that these are for preliminary layout only in well-ventilated locations.

Definitions

Liquid = fluid below atm. b.p. (see Table A3.6 for definitions of Class A, B and C fluids)
Gas = fluid above atm. b.p.

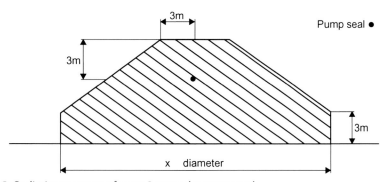

Figure A3.3 Preliminary extent of zone 2 around a pump seal.

Table A3.9 Electrical area classification distances for centrifugal pumps

Seal	Fluid conditions	Zone 1	Zone 2 X in Fig A3.3
Any (inc. reciprocating pumps)	Liquid $<$ atm. b.p. $\geq$ ambient temperature	None	Diameter of pool $+$ 6 m
Mechanical seal, external throttle bush, drain, atm. b.p. $>$ ambient temperature No bush need be used for cases marked * if X doubled	Class A, $>$ atm. b.p. temp $<100°$c temp $<200°$c temp $>200°$c Class B, $>$ atm. b.p. temp $<200°$c temp $>200°$c Class C, $>$ atm. b.p. temp $<250°$c temp $>250°$c	0.3 m sphere around seal	20* 40 60 20* 50 As liquid 20
Mechanical seal, external throttle bush, vent to stack, atm. b.p. $\leq$ ambient temperature	Liquefied C_4's (i.e. atm. b.p. $\approx 0°$c) Liquefied C_3's and lighter HC (i.e. atm. b.p. $\approx 20°$c) (see note below) Liquefied non-hydrocarbons	0.3 m sphere around seal	40 60 20–30

Note: Zone 1 for C_3's which may be up to 3 m depending on seal performance.

Table A3.10 Electrical area classification distances for equipment other than pumps

Item	Condition	Zone 1	Zone 2
Compressors in open-sided houses	Gases	See note below	See Fig A3.4

Note: Zone 1 is 0.5 m around any gland, seal, drain parts, vents except 1 m is allowed around a seal oil lid and vent or a seal oil trap

Equipment in normal buildings		Outdoor distances as shown in Fig A3.5	
Joints and flanges on pipes, fittings and process equipment	Liquid Gas lighter than air Gas heavier than air	None None None	X = diameter of pool $+$ 6 m in Fig A3.3 3 m horizontal status, 7.5 m above, 5 m below 7.5 m horizontal radius, 5 m above and down to floor

Note: Valve glands can be treated as pump seals

(Continued)

Table A3.10 (Continued)

Item	Condition	Zone 1	Zone 2
Relief valves, vents etc.	High velocity, gas lighter than air	1 m sphere	See Fig A3.6 H = 100, R = 60
	High velocity, gas heavier than air	1 m sphere	See Fig A3.6 H = 260, R = 120
	Low velocity, frequent release	1.5 m sphere	3 m sphere
	Low velocity, infrequent release	None	
Sample points <6 mm diameter	Liquids near ambient temperature, into open	None	See Fig A3.7
	Other liquids into closed system	None	15 m radius, 3 m up, down to floor
	Gases into closed system	None	See 'Joints, and flanges on pipes etc.' above
Process water drain point into open, at grade used regularly	Liquids	See note below	X = diameter of pool + 6 m in Fig A3.3
	C₃ under pressure	See note below	3 m high × 45 m radius
	C₄ under pressure	See note below	3 m high × 30 m radius
	Other gases under pressure		3 m high × 20 m radius

Note: Zone 1 is a cylinder 1 m radius and 1.5 m high for liquids and 5 m radius and 1.5 m high for gases

Item	Condition	Zone 1	Zone 2
Instruments etc., near or at grade	Liquids	See note below	X = diameter of pool + 6 m in Fig A3.3
	Gases	See note below	Flanges as pipe joints Drains as sample points

Note: Zone 1 is not needed for infrequent spills but otherwise is a cylinder 3 m high by radius of 3 m if below atm. b. p. and 5 m if above atm. b. p.

Item	Condition	Zone 1	Zone 2
Road or rail (un) loading	Liquids	See Fig A3.8 for zones 1 and 2; H = 1 m	
	Gases	See Fig A3.8 for zones 1 and 2; H = 3 m	
Ship (un)loading		20 m around × ∞ high	None

(Continued)

Table A3.10 (Continued)

Item	Condition	Zone 1	Zone 2
Unloading only		None	20 m around except seaside × 10 m high
Fixed roof tank	Liquids	See Fig A3.9 for zones 0−2 See also Preliminary Spacings for Tank Farm Layout	
Floating roof tank	Liquids	See Fig A3.10 for zones 0−2 See also Preliminary Spacings for Tank Farm Layout	
Pressure storage vessel	Gases	See 'Joints and flanges on pipes, etc.', 'Relief valves', 'Process water drain point', as appropriate	
Low pressure refrigerated tank		See also Preliminary Spacings for Tank Farm Layout	
Open topped oil water separator	Liquids	See Fig A3.11 for zones 0−2	
Open topped drains and effluent pits	Liquids	See Fig A3.11 for zones 1−2	
Drums in open	Liquids	See Fig A3.12 (only if being filled)	3 m around drum area

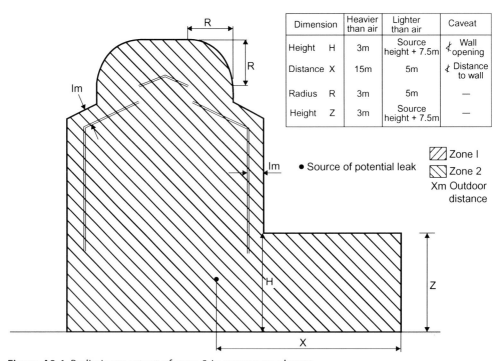

Dimension		Heavier than air	Lighter than air	Caveat
Height	H	3m	Source height + 7.5m	∡ Wall opening
Distance	X	15m	5m	∡ Distance to wall
Radius	R	3m	5m	—
Height	Z	3m	Source height + 7.5m	—

● Source of potential leak

⧄ Zone l
⧄ Zone 2
Xm Outdoor distance

Figure A3.4 Preliminary extent of zone 2 in compressor house.

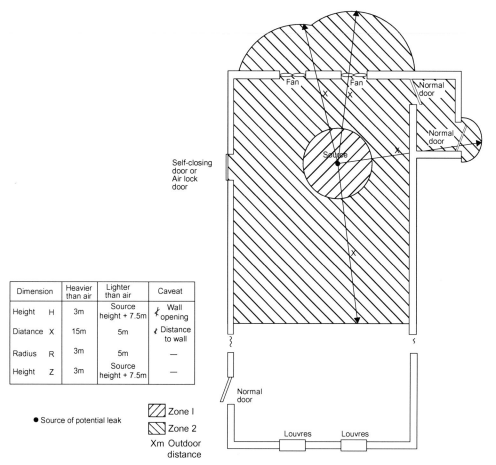

Dimension		Heavier than air	Lighter than air	Caveat
Height	H	3m	Source height + 7.5m	⚡ Wall opening
Diatance	X	15m	5m	⚡ Distance to wall
Radius	R	3m	5m	—
Height	Z	3m	Source height + 7.5m	—

● Source of potential leak

▨ Zone 1
◪ Zone 2
Xm Outdoor distance

Figure A3.5 Preliminary extension of zones to outside building.

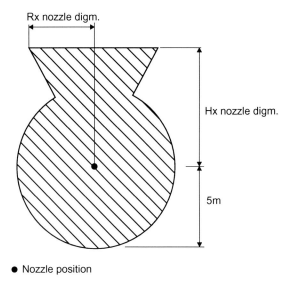

Rx nozzle digm.

Hx nozzle digm.

5m

● Nozzle position

Figure A3.6 Preliminary extent of zone 2 around a relief valve, etc.

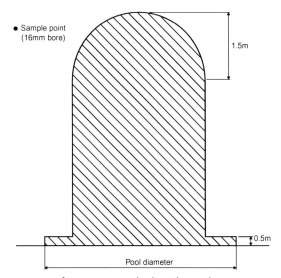

● Sample point
(16mm bore)

1.5m

0.5m

Pool diameter

Figure A3.7 Preliminary extent of zone 2 around a liquid sample point.

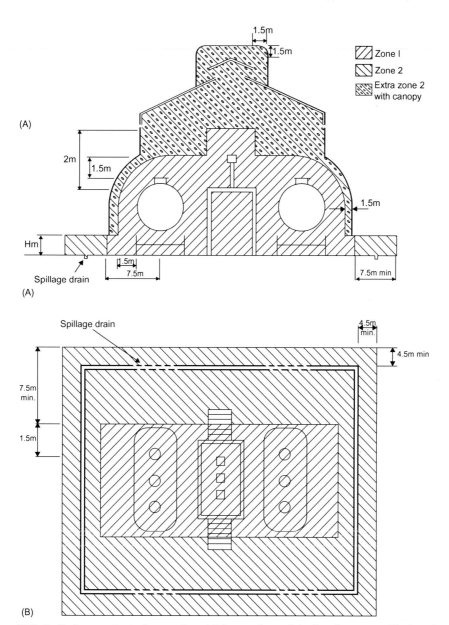

Figure A3.8 Preliminary extent of zones 1 and 2 for road or rail (un)loading areas: (A) elevation; (B) plan view.

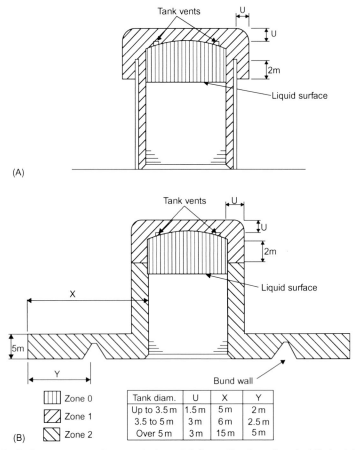

Figure A3.9 Preliminary extent of zones 0, 1, and 2 for a fixed roof tank: (A) double-walled tank; (B) single-walled tank.

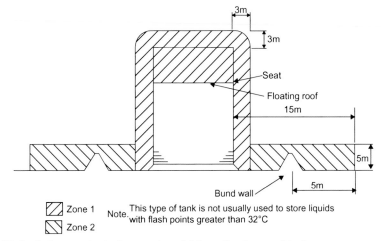

Figure A3.10 Preliminary extent of zones 1 and 2 for a floating roof tank.

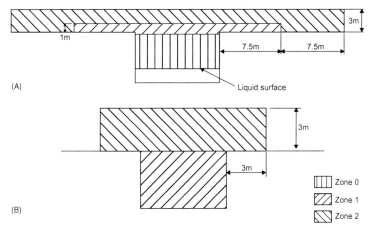

Figure A3.11 Preliminary extent of zones 0, 1, and 2 in open-topped constructions: (A) open-topped oil/water separator; (B) quench drain channel or effluent interceptor pit.

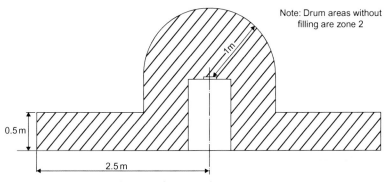

Figure A3.12 Preliminary extent of zone 1 for drum filling in open.

SIZE OF STORAGE PILES

1. The height (h, in m) of a right conical pile is given by:

$$h = \left(\frac{3V\tan^2 O}{\pi}\right)^{1/3}$$

where $V =$ volume (m^3) and $O =$ angle of repose (commonly 37°). If the conveyor angle is ϕ, the horizontal length of the conveyor (L_1 in m) is $L_1 = h \cot \phi$. The angle ϕ is commonly 18°. The radius (r, in m) of the bottom of the pile is $r = h \cot O$. It follows that the minimum length (L_2 in m) required for a conveyor and pile in one straight line on plan is:

$$L_2 = L_1 + r = h(\cot \phi + \cot O)$$

2. Approximate volume (V, in m$_3$) of a straight conical pile is:

$$V = h^2 L_3 \cot O + \frac{\pi}{3} h^3 \cot^2 O$$

Where L_3 (in m) is the length of the top of the pile.

3. Approximate volume (V, in m^3) of a curved conical pile is:

$$V = h^2 R\alpha \cot \theta + \frac{\pi}{3} h^3 \cot^2 \theta$$

Where $R =$ radius of curve (in m) and $\alpha =$ size of arc in radians

4. Approximate volume of closed warehouse (V, in m^3) is:

$$V = h^2 L_4 \cot \theta$$

Where L_4 (in m) is the length of the pile. This equation assumes fully triangular cross-section and no spaces around the piles for conveyors or mechanical unloading equipment. Thus the equation can be used as it is for underground conveying, but for unloading from one side, add 5 m to the width of the store. Also add 10−20% to the length to allow for dead spaces.

APPENDIX 4

Checklists for Engineering Flow Diagrams

This appendix is reproduced and adapted from Sandler, H.J. and Luckiewicz, E.T. (1987) *Practical Process Engineering: A Working Approach to Plant Design*.

The placement of the required equipment and necessary connecting piping with identifications on an engineering flow diagram (such as a PFD or P&ID) is only the initial phase of the work required to make the diagram complete. A great many details must be added so that the process being described on the diagram fulfils its function in an economical and safe manner and is in compliance with environmental requirements. While the details entered vary considerably from project to project, many items are common to most projects. The following checklists regarding major components found on many engineering diagrams should be kept in mind while preparing the diagrams. The lists concern the appurtenances on equipment and piping and also review primary instrumentation as well as important safety considerations.

Although the lists are not fully comprehensive, the items in them relate to situations most likely to be encountered and alert the process engineer to the type of details to be considered for inclusion on engineering flow diagrams.

Short elaborations of the various items in the checklists follow to give guidance for their representation on an engineering flow diagram.

Table A4.1 Checklists for engineering flow diagrams

Type of equipment	Item	Notes
Vessels	Nozzle types	Connections:
		• Couplings, generally for connections of 32mm nominal bore (NB) diameter or smaller.
		• Flanged nozzles, for any size of connection. Nozzles with a diameter smaller than 50mm NB on low-pressure vessels are fragile, and connections should be made by means of reducing fittings or flanges.
		• Non-metallic connections are often of limited size. Manufacturer catalogues should be consulted.
	Vents	• Closed vessels are generally vented.
		• Vents with only air or low amounts of water vapor terminate above the top of the vessel.
		• Vents with vapors that may be harmful but are not toxic or lethal (i.e. hot gases or hot vapors) may terminate outdoors above adjacent structures in accordance with pertinent regulations.
		• Dangerous vapors or gases pass to a flare system or to a collection system for further treatment.
		• An invert or 'rain hat' protects vessel contents from precipitation.
		• The diameter of the vent line is generally made equal at least to that of the largest liquid line entering the vessel.
	Drains	• Most vessels are provided with drains.
		• Drain connections can originate from a discharge line or from the vessel itself.
		• The destination of a drain should be indicated.
	Vortex breakers	• A vessel with a low liquid level may require a vortex breaker to prevent gases from entering a pump suction.
	Overflow pipes	• Open or low-pressure vessels handling liquids require an overflow connection.
		• The top invert of the overflow nozzle must be sufficiently below the top tangent line to accommodate the constriction head at the maximum flow rate through the nozzle.
		• Overflow lines for closed, blanketed vessels or those under a slight negative pressure must be sealed in a suitable liquid loop seal or by means of a mechanical sealing trap.
		• Seal-leg heights should be included on the engineering flow diagram if required to prevent over pressuring or collapse of the vessel.
		• A high-level switch can be used to close valves in inlet lines or to stop supply pumps.

(Continued)

Table A4.1 (Continued)

Type of equipment	Item	Notes
	Sample connections	• Sample connections may be placed on a vessel, on a discharge line at the bottom of the vessel, or on a continuously flowing line from a pump taking suction from the vessel. • Samples may be taken from the top of a vessel through a suitable sampling mechanism. • A series of connections is required for vessels which may have stratified layers.
	Dip legs	• Dip legs are required to diminish the generation of static-electricity effects with flammable liquids, reduce foaming, introduce reactants into an agitated vessel, return heated, viscous fluids to the suction nozzle of a circulating pump, create a barometric leg in a vacuum system, or prevent corrosive effects due to aeration. • A weep hole or a siphon break is frequently required in dip legs other than those acting as barometric legs. • Weep holes in lines with vapor-liquid flow should be of minimum size and located in the neck of the nozzle.
	Gooseneck inlets	• A gooseneck inlet is an alternative to a dip leg in a metallic vessel to reduce static-electricity effects. • A weep hole is not required with a gooseneck inlet.
	Baffles	• Baffles are frequently required to improve agitation. • The vendor of agitation equipment usually recommends the baffle sizing and configuration best suited for the respective vessel dimensions.
	Manholes	• Manholes are required for inspections, cleanout and maintenance. • For ventilation and as an escape or rescue port, one manhole should be in the head or top of a vessel with another on the side or bottom. • Manholes are normally of 600mm diameter but never smaller than a 400mm × 600mm ellipsoid for use in special instances. • Unwanted dead space in a manhole can be avoided by use of an internal 'hat' or a curved flush cover. • Number, sizes and special configurations are to be shown on engineering flow diagrams.

(Continued)

Table A4.1 (Continued)

Type of equipment	Item	Notes
	Handholds	• Handholds are used on small vessels for inspecting limited areas, feeling for the integrity of a vessel, and charging or removing catalysts, desiccants, or packing from columns or vessels. • Number, sizes and special configurations are to be called out on flow diagrams.
	Air locks	• An air lock is a small vessel mounted at the top or directly beneath a bin or reactor to introduce or remove solids or liquid intermittently. • Suitable valves are placed upstream and downstream on the air lock. One or both valves are always closed to isolate the bin or reactor from its surroundings. • An inert purge may be used to preclude air or water vapor from a bin or reactor or to prevent the atmosphere in the vessel from escaping to the surroundings.
	Rotary feeders	• A rotary or star feeder permits a continuous, controlled, adjustable flow of solids into or from a vessel while isolating it from its surroundings. • An inert purge may be used to preclude air or water vapor from a vessel or to prevent the atmosphere in the vessel from escaping to the surroundings.
	Live bottoms	• The containing vessel should have a coned bottom whose angle is greater than the angle of repose of the solids. • An internal pair of small cones, one inverted from the other, facilitates the outflow of solids. • An air lance is frequently used to prevent or disrupt bridging. • One or more mechanical vibrators attached to the outside of the cone bottom helps to induce the flow of solids. • A 'live bottom', or pulsating section, is often inserted between the cone section and the discharge valve for difficult solids-flow problems.
	Vessel jackets	• Carbon steel vessel jackets offer a low-cost means of transferring heat to glass-lined or high-alloy vessels. • Liquid media enter the bottom of a jacket and follow a tortuous path through the baffled or dimpled annulus to exit at the top. • Condensing media enter at the top, and the condensate leaves at the bottom; baffling is not needed. A vent should be provided.

(Continued)

Table A4.1 (Continued)

Type of equipment	Item	Notes
	Heat transfer panels	• Steam and water may be used alternately, but vents and drains must be provided in utility piping so that the two media do not contact one another. Controls should be provided in utility piping to glass-lined tanks to prevent high temperature differentials between the vessel and the jacket contents. • The panels are used to transfer modest amounts of heat or in retrofitting a vessel. They are frequently used to maintain the temperature of vessel contents against heat loss or gain. • External panels may be constructed of carbon steel or galvanized sheet regardless of the vessel material. • They are available in a variety of shapes to fit various contours. They may be supplied with different internal paths, serpentine configurations, and positions for inlet and discharge connections. • Usually they are given item numbers separate from that of the vessel.
	Internal coils	• These coils are placed inside agitated vessels for improved heat transfer. They may be used by themselves or in addition to a jacket. • They may consist of more than one concentric set of coils provided there is adequate space for circulation between coils and rows. • The materials of construction must be compatible with the contents of the vessel at the extremes of temperature expected within the coil.
	Vessel tracing	• Heat needed to hold a storage vessel at a required temperature or to provide freeze protection is sometimes supplied by tracing the vessel with electrical tape. Heat transfer is enhanced by the use of a special grout or mastic. • Sheets of metallic foil can be used to cover vessels constructed of plastic, resinous, or other non-metallic materials before the application of electric-tracing tape to prevent the generation of localized hot spots. • The presence of electric tracing on a vessel is normally not indicated in the body of an engineering flow diagram but is noted with the item number, name, and description.

(Continued)

Table A4.1 (Continued)

Type of equipment	Item	Notes
	Insulation	• Some design groups indicate insulation by portraying a section of it on a vessel in the body of the flow diagram; others include a notation 'Insulate' or the purpose of the insulation in the section with the item description.
	Removable spool pieces	• Vessels which personnel can enter should be provided with removable spool pieces at liquid inlet nozzles servicing the vessel. • Spool pieces are removed and piping or valves blanked off whenever a person is in the vessel.
	Vessel position	• Engineering flow diagrams should show whether vessels are in a vertical, horizontal or inclined position.
	Vessel supports	• Representations of vessel supports should be included in flow diagrams.
Agitators or Mixers	Location of agitators	• Most agitators enter at the top, and the shaft extends straight down or at an angle. • A side-entering agitator is mounted through a nozzle below the liquid level. • In some instances, the shaft may enter through the bottom head.
	Types of agitators	• The type of agitator (propeller, turbine, anchor etc.) should be depicted on the flow diagram. • Multiple sets of blades are sometimes required and should be shown. • Auxiliary internals such as vessel baffles or draft tubes are often associated with agitators. • In the absence of prior similar or pilot-plant experience, a reputable agitator or mixer manufacturer should be consulted to determine the type of agitator or auxiliaries.
	Steady bearings	• A steady bearing should be denoted on a flow diagram if one is included to reduce the size of the shaft diameter.
	Seals	• Packing or a mechanical seal is used to prevent leakage of air into or vapor out of a closed vessel. • Shaft packing or a mechanical seal is required for a side- or bottom-entering agitator. • While shaft packing or mechanical seals generally are not noted on engineering flow diagrams, special lubrication or cooling systems are depicted.

(Continued)

Table A4.1 (Continued)

Type of equipment	Item	Notes
Pumps	Types of pumps	• A large variety of pumps is available for use in a processing plant (e.g. centrifugal ump, reciprocal pump, gear pump, metering pump, diaphragm pump (air-operated), in-line pump).
	Types of drivers	• Since most pumps are driven by electric motors, many design groups usually omit a driver symbol if the driver is an electric motor.
		• An air motor may be employed if the power requirement is low or for safety considerations in electrically hazardous areas.
		• Hydraulic motors are used for high-torque requirements in special applications.
		• Steam turbines are used as an economy measure when exhaust steam is available. They are also used for critical pumps or compressors when flow is to be maintained during a power failure.
		• Diesel and gasoline engines are used as alternatives to steam for critical services. They are also used as prime movers in remote locations.
	Valved vents and drains	• Vents and drains can be provided on a pump casing rather than on the discharge or suction piping.
		• The type and size of valve are to be shown at the appropriate location on the pump symbol.
		• T-bar handles can be welded to casing plugs. They should be indicated on the flow diagram.
	Quench system	• Pump packings or seals must be protected from high-temperature fluids by a suitable cooling-water quench.
		• Some types of seals require special jacketing.
		• Water piping to and from pump quench or seal jackets is shown on the flow diagram.
	Flushing and seal fluid systems	• Water or other compatible fluid is sometimes required to flush slurries, crystallizing solutions, or corrosive liquids from shaft packings.
		• Liquids may be needed to flush a mechanical seal to remove frictional heat or to prevent a slurry from reaching a seal face.
	Jacketed pumps	• Jacketed casings are used to maintain the fluid in the pump at an elevated or a chilled temperature.
		• Jacketing is also used to protect the pump casing from excessive temperatures.

(Continued)

Table A4.1 (Continued)

Type of equipment	Item	Notes
Heat Exchangers	Tracing and insulation	• Heat tracing may be required to prevent residual fluid in a pump from freezing when the pump is not operating. • Insulation may be required over tracing or jackets to conserve heat or energy. It may also be needed for personnel protection or antisweat purposes. • Tracing and insulation requirements are included with the equipment description.
	Pump recycle lines	• A continuous recycle line is sometimes installed to maintain minimum flow through a pump. • A restriction orifice is normally placed in the recycle line.
	Types and configurations	• The representation of the heat exchanger on the flow diagram should reflect as much as possible its particular type and configuration. • The location of the nozzles should approximate those on the exchanger as it is to be placed in service.
	Sloping exchangers	• Single-tube-pass horizontal exchangers in condensing service are frequently sloped in the direction of the condensate outlet.
	Position of nozzles	• Liquids being heated should enter exchangers at the low point and leave at a high point to prevent buildup of air or other gases which may come out of solution. • If suspended solids are present, the flow should be from the top down to prevent a buildup of solids. • Gases may be removed from a two-shell-pass exchanger by an external restricted vent or from the channel of a vertical U-tube exchanger by an internal weep hole. The orifices are sized to preclude excessive bypassing of fluid.
	Backflushing exchangers	• Periodic backflushing is needed if a fluid contains suspended matter and there is a wide variation in its velocity as it passes through the exchanger. • Cross piping and valving are provided to direct normal inlet fluid to the discharge nozzle while effluent leaves the inlet nozzle carrying dislodged suspended material to the discharge piping.

(Continued)

Table A4.1 (Continued)

Type of equipment	Item	Notes
	Impingement protection	• In many exchangers, it is necessary to protect the tubes or internals from erosion by high-velocity fluids entering the shell. The exchanger manufacturer should be consulted to determine whether or not such protection is required and what form it should have. • Protection is afforded by an impingement baffle beyond the fluid inlet nozzle, a dome at the inlet nozzle, or an oversized nozzle.
	Vents and drains	• Vents and drains, when removed, are usually placed in the inlet or discharge piping to an exchanger; however, they can be made to connections on the inlet and outlet discharge nozzles or to a boring through the tubesheet.
	Tracing and insulation	• Tracing and insulation are required to prevent one or both sides from freezing unless a small continuous flow is sufficient to prevent freezing or it is permissible to drain the affected side of the exchanger when it is not in use. • Insulation may be required for heat or energy conservation, personnel protection, or antisweat reasons. • Requirements for insulation and tracing are indicated on a flow diagram as part of the equipment description. • It is not feasible to trace or insulate some exchangers such as plate-type units. Shrouds must be indicated for them for personnel protection, and they must be drained when not in use to prevent freezing.
Fans, blowers and compressors	Gas movers	• An appropriate symbol should be entered on the flow diagram.
	Types of drivers	• Gas movers are usually driven by electric motors. • Compressors sometimes are operated by steam turbines if excess steam is available or for emergency use during an electrical interruption. • Gasoline or diesel engines are employed for emergency conditions or at remote locations.
	Coolers	• The heat developed in a compressor must be removed if the discharge temperature exceeds the unit's maximum allowable value, usually about 177°C.

(Continued)

Table A4.1 (Continued)

Type of equipment	Item	Notes
		• Some compressors are designed to circulate cooling water in jackets or internal passages. Filtration is often required to prevent clogging if municipal water is not used as the cooling medium.
		• Intercoolers are frequently used between compressor stages with an aftercooler following the unit; provision is made to remove condensate.
		• Large compressors usually have separate lubrication systems including external lubricating-oil coolers which require municipal and filtered cooling water.
	Knockout pots	• These are required in the suction line to a blower or compressor if there is liquid carryover from the preceding step of the process or if condensate can form in the suction piping.
	Valved drains	• Despite the use of knockout pots, moisture may enter a blower or compressor and coalesce.
		• Drains also are needed to remove wash fluid that may be injected intermittently to clean the blower or compressor blades.
	Accumulators	• Accumulators are needed to smooth suction gas flow if the upstream volume is relatively small or if there are upstream pressure variations.
		• They are needed to smooth pulsations from a reciprocating compressor and reduce downstream pressure variation.
	Screens, filters and silencers	• A screen may be needed in the inlet line of an air blower or compressor to prevent foreign objects from entering the unit.
		• A filter may be required if suction gas contains small solid particles which would cause excessive wear or a buildup of deposits. Non-lubricating reciprocating compressors are particularly subject to the abrasion of piston rings.
		• Filters are used after oil knockout pots on lubricated compressors to remove fine oil mists.
		• Silencers are sometimes needed on the suction or discharge sides of blowers or compressors to reduce noise to an acceptable level. Some high-speed centrifugals must be provided with a noise-abatement envelope.

(*Continued*)

Table A4.1 (Continued)

Type of equipment	Item	Notes
	Controls and interlocks	• Primary sensors and control valves are shown on engineering flow diagrams to ensure economical and satisfactory operation of blowers and compressors and proper functioning of auxiliary cooling and lubrication systems.
Vacuum Equipment	Types of vacuum equipment Drives and motive forces	• The particular type of vacuum equipment should be represented by an appropriate symbol. • Mechanical types of vacuum equipment are usually operated by electric motors. • Vacuum equipment with non-moving parts is powered by a high-pressure fluid, usually steam, air, or water. Piping for the motivating fluid is shown on the flow diagram.
	Cooling vacuum equipment	• The heat of compression is removed from non-wetted mechanical equipment by circulating water through jackets or internal passages. Interstage cooling is required if there is more than one pumping stage. • Liquid-ring vacuum pumps are cooled by a modest flow of fresh water or a cooled recycle and compatible sealant fluid. There is usually a sufficient pressure differential between the discharge and suction sides to accomplish the recycle through an exchanger. • Steam-jet ejector systems are supplied with interstage condensers once the interstage dew-point temperature is above that of the cooling medium. Condensation may take place either by direct contact with the cooling medium or indirectly in a heat exchanger. • Condensate from an interstage condenser is subatmospheric and is collected through a closed dip leg in a hot well or is pumped to disposal. The flow diagram should include a note regarding the minimum height of the interstage condenser above the hot well, taking into account the specific gravity of the continuous phase of the condensate.
	Knockout pots and traps	• Inlet knockout pots are provided to prevent entrained liquids from entering rotary, screw or reciprocating-type vacuum pumps.

(Continued)

Table A4.1 (Continued)

Type of equipment	Item	Notes
		• Oil-lubricated units are followed by a knockout pot to recover and recycle oil. Knockout pots follow liquid-ring vacuum pumps to separate the sealant from the discharge gas. • Motive-steam lines to jet ejectors should contain a condensate separator and be well trapped to prevent intermittent condensate particles from damaging the orifices in the jets.
Filters and centrifuges	Types of filters and centrifuges	• Representations of filters and centrifuges in an engineering flow diagram should be typical of their configuration.
	Types of drivers	• Centrifuges and rotating filters usually are powered by electric motors; some centrifuges use hydraulic motors to drive solids-removing plows. • Compressed air or nitrogen is frequently used to dislodge solids from filter socks in a baghouse or to inflate flexible members in some plate-and-frame filters for the compression of the cake or to assist in its discharge.
	Additives and precoating	• Some liquid-solids separations require the addition of surfactants or coagulants before filtration. • It is often necessary to precoat a filter or to mix filter aid with the filter feed. • All mix tanks, agitators, pumps, and controls are to be shown on the flow diagrams as auxiliaries to the filtration.
	Recycle and sampling	• A recycle line is included in batch operations to ensure sufficient buildup of cake or filter medium to give the desired clarity. • A sampling point should be provided in a filtrate discharge line if sampling ports are not provided on the filter.
	Auxiliary lines and equipment	• A wash system is frequently required for filters and centrifuges. • A means to contain and transfer segregated solids is needed. • Provision must be made to collect and transfer mother liquor from a centrifuge. • Drum filters require vacuum equipment as an auxiliary. • Piping and equipment for all auxiliaries are shown on the flow diagrams.

(Continued)

Table A4.1 (Continued)

Type of equipment	Item	Notes
Piping		A considerable amount of the work represented on engineering flow diagrams is concerned with piping. The following are several important aspects which should be kept in mind when preparing the diagrams.
	Valves	• Valves are used to isolate piping or equipment from other portions of the process and to throttle or divert flows; thus, they are an essential part of any piping system. Valves should be designated on engineering flow diagrams by a symbol which not only shows where a valve is required but also indicates the type of valve.
	Bypasses for control valves	• On-off valves, i.e. blocking valves, are usually placed immediately upstream and downstream of an automated control valve in continuous or critical service along with a bypass line containing a manual throttling valve about the group. Such configuration permits continued operation or orderly and safe shutdown of a process in the event of a controller or control-valve malfunction.
		• The block valves and bypass line and valve are usually of line size for main lines up to 75mm NB in diameter. For 75mm NB main lines and larger, it is good practice to have the bypass line and valve one line size smaller.
		• Some self-regulating valves in clean service, such as air-pressure regulators, are seldom provided with a bypass line.
	Vent and drain sizes	• Vent lines on a vessel should be of at least the size of the largest incoming recycled line.
	Condensate traps	• Vent lines or relief-valve discharge lines which carry steam or condensable water vapor and have a horizontal section that can form a seal leg should be provided with a drain hole, if permissible, or with a suitable condensate trap at low points to prevent excessive pressure drops or the development of excessive and dangerous velocities of trapped liquids.
		• A manual drain is usually suitable for intermittent operations.
	Protective screens	• Atmospheric lines to the suction of a blower or compressor as well as pump intake lines from natural bodies of water or from sumps should be protected by screens from the entrance of foreign bodies.

(Continued)

Table A4.1 (Continued)

Type of equipment	Item	Notes
	Jacketing and bundling	• An outer pipe, known as a jacket, is sometimes placed around a process line. An appropriate heating or cooling utility stream flows in the annulus to maintain the process fluid at a desired temperature. • Jacketing is shown on the body of the flow diagram, and the word 'jacketed' is entered in the tracing column on the line tabulation. • A line is 'bundled' when it is placed next to a hot or refrigerated line, and both lines are then insulated together. The word 'bundled' is placed in the tracing column of the line tabulation.
	Boundary definitions	• While building outlines per se are seldom defined on engineering flow diagrams, it may be necessary to locate a change point through which a pipeline passes for tracing, insulation, piping classification, or safety reasons. • The transition between classifications is marked by a short indicator perpendicular to the piping line. The boundary definitions are delineated.
Instrumentation and Safety	The process engineer enters all sensing instrumentation, control valves, and safety devices on the P&ID or engineering flow diagram and designates whether instrument measurements are to local, or remotely indicated, or as recorded points. The instrument engineer completes the control loops in the P&ID or on a separate instrumentation flow diagram. The following subsections discuss some of the more common instruments and safety devices to be considered for application on a flow diagram. Other specialized items should be provided as required.	

	Relief valves	• A relief valve, also known as a pressure safety valve, protects equipment or piping from pressures beyond its design maximum allowable working pressure (MAWP).
		• Typical situations for relief-valve application are:
		  ○ Any vessel, exchanger, column or other equipment that can be completely isolated by valving must be protected from an external fire or runaway exothermic heats of reaction.
		  ○ Vessels with an open vent or overflow that is of such size or length that excessive pressure could be generated in the event of fire or reaction excursion must be protected.

(Continued)

Table A4.1 (Continued)

Type of equipment	Item	Notes
		○ Protection is also needed when the maximum discharge pressure of a pump or compressor feeding a piece of equipment or piping is greater than its MWAP and the pump or compressor can be deadheaded through the equipment or piping. ○ Exchangers need protection when a liquid 'cold' side is isolated and expanded owing to the continued flow of the 'hot' side. ○ A relief valve is usually placed after a pressure-reduction valve to ensure that subsequent equipment is protected in the event of a malfunction of the reduction valve. ○ If a relief valve may not reseat completely owing to fouling by solids or gummy materials, a pair of relief valves is used in parallel and two ganged three-way valves are incorporated in the piping so that there is always one relief valve functioning with the equipment while the other is being cleaned. • The sizes of relief-valve inlet and discharge lines are shown on the flow diagrams. The discharge side of most vapor safety valves is usually larger than that of the inlet, while those in liquid service usually have equal inlet and discharge connections.
	Rupture disks	• A rupture disk is an alternative to a relief valve when it is acceptable to allow pressure in equipment or piping to fall and to lose material until atmospheric pressure is reached. • It consists of a frangible wafer of composite materials or a thin metallic element which is shattered or ripped apart at a predetermined pressure. • The material of the wafer and the metallic element must be compatible with the fluid in the equipment or piping. A thin membrane of Teflon or other polymeric material is used to prevent chemical attack of the disk and results in an economical construction.

(Continued)

Table A4.1 (Continued)

Type of equipment	Item	Notes
	Vacuum breakers	• A rupture disk can be placed ahead of a relief valve to protect the relief valve from plugging or to permit the valve to be constructed of more economical materials. A pressure gauge is usually placed between the relief valve and the rupture disk to indicate the integrity of the disk. • A vacuum breaker permits atmospheric air or a compatible gas or vapor to enter a vessel at a determined vacuum level to prevent collapse of the unit at its maximum design pressure. • Typical situations that require a vacuum breaker are the withdrawal of liquid with insufficient replacement of gas or vapor, the condensation of vapors in an isolated piece of equipment whereby the pressure in the vessel is reduced, and the connection of vessels to a powered vent system.
	Conservation vents	• A conservation vent is a combination of a relief valve and a vacuum breaker. • It normally allows storage vessels containing volatile fluids to float within a limited pressure range so that the loss of vapors is reduced.
	Flame arresters	• The vent on any unit containing an inflammable fluid should be provided with a flame arrester to prevent backflushing if the discharge vapors become ignited. • Conservation vents may be purchased with integral flame arresters.
	Pressure sensors	• Pressure connections to equipment should be made, where possible, in the gas or vapor space to eliminate corrections for hydraulic head. • Units such as filters, baghouses, heat exchangers, or distillation columns which induce considerable pressure drops often have a differential-pressure measurement across the unit in addition to an absolute- or gauge-pressure sensor at the inlet or discharge. • A pressure gauge on the discharge of a pump provides information regarding the operation of the pump; a permanent gauge (PI) may be installed, or a valved provisional tap (PP) may be used instead.

(Continued)

Table A4.1 (Continued)

Type of equipment	Item	Notes
	Temperature sensors	• Pressure gauges should be isolated from equipment or piping with a suitable valve or chemical seal. The latter is used with corrosive fluids, toxic fluids, slurries, or fluids that would solidify in the gauge. All isolating valves and chemical seals are shown on the flow diagrams. • Whenever applicable, thermowells are located in the liquid portion of a two-phase system to give quicker responses. • Thermowells, with or without indicators, are frequently provided upstream and downstream of a heat exchanger. • It is good practice to provide a redundant temperature measurement to activate an alarm function in addition to the normal temperature measurement used to control an exothermic reaction.
	Sight glasses and lights	• A viewing port is often included in agitated vessels or on distillation columns especially if there may be foaming problems. • If caustic or hydrogen fluoride is present, glass ports must be protected by a suitable transparent insert between them and the fluid. • Sight glasses are available in multiple sizes and configurations for use in piping. They are employed to indicate the presence of liquid or an interface, whether the liquid is stationary or flowing (the turning of a wheel), and the approximate flow rate (the lifting of a flapper). • A sight glass is often accompanied by a sight light; all sight glasses and lights are shown on the flow diagrams.
	Level indicators	• All process vessels containing liquids require level measurements. • The simplest indicators are a shadowed liquid level through the translucent wall of a plastic tank with a calibration on the outside of the tank or a calibrated dipstick for use as an intermittent indicator in non-critical, non-hazardous service. • A simple continuous-indicating device is a window or a series of windows in the straight side or a vertical cylindrical glass tube along the side of a vessel. A series of tubes is required to determine an interfacial level.

(Continued)

Table A4.1 (Continued)

Type of equipment	Item	Notes
		• A series of hollow external metal columns provided with translucent reflux faces is used when the application of glass windows or cylinders is limited by the design pressure of the vessel. This system is not suitable for showing interfaces. • Means must be provided to drain, flush and clean external gauges. • A pressure-sensing device may be used at or near the bottom of the vessel as an indication of level. A differential-pressure unit with one leg connected to the vapor space is required if the tank is not open or if its pressure in the vapor space is greater than a few inches of water column. • A chemical seal is used to isolate the pressure sensor from liquids which are corrosive or toxic, could freeze at ambient temperatures, or are slurries. • A common level indicator uses a top-entering dip tube through which a small metered flow of compressed air, nitrogen or other compatible gas is metered and bubbles trough the liquid. The pressure of the gas, corrected for liquid density, indicates the height of the liquid above the end of the dip tube. A differential-type pressure indicator is required for pressurized vessels. • Bubble-type indicators should not be used with saturated or nearly saturated solutions since evaporation may leave deposits at the end of the tube and lead to obstructions, causing erroneous level readings. • A guided float sensor gives a continuous level reading of the surface of a liquid or a pile of solids. It is also used to indicate a narrow-gauge 'either-or' local condition. • Paddles, flappers, capacitance and sonic probes, and radiation units are some of the devices employed for specialized level indication. Strain gauges are used to determine weight. Notes on the flow diagrams are used to denote these special level indicators.

(Continued)

Table A4.1 (Continued)

Type of equipment	Item	Notes
	Flexible connections	• Equipment or piping made of ceramics, glass and certain resinous materials is isolated by flexible connections or hoses from damaging vibrations caused by mechanical equipment, especially reciprocating or high-speed types of gas movers. • Flexible connectors are frequently used to ensure that liquid lines of 200mm NB diameter or larger do not overstress inlet or discharge nozzles on pumps, filters, or similar equipment. • Flexible connections serve to isolate a weigh tank from its fixed entrance, discharge, or vent lines to permit unimpeded movement of the vessel.
	Expansion joints	• An expansion joint is used in piping in lieu of bends and loops to account for linear growth due to changes in temperature. • Differential linear growth between the two sides of a heat exchanger is usually relieved by placing an expansion joint in the shell unless the tubes are coils or U tubes or the unit has a floating, pull-through, or packed-head configuration. • Consideration should be given to the extremes of start-up or emergency conditions as well as normal operating temperature in determining whether or not an expansion joint is required.
	Fire valves	• These are ball or plug valves which maintain their closure by a built-in spring mechanism but are held open in normal operation by a link mechanism containing a fusible section. • They are recommended for installation at draw-off nozzles on vessels with flammable contents.
	Spring-closure valves	• These valves are used when it is important that a manually operated valve returns to a closed, open, or partially open position after being used. • They are often used instead of a restriction orifice in services which tend to clog openings. If the valve begins to clog, it can be opened fully and then returned to a stop for its partially open position.
	Locked valves	• Certain valves should remain in an open or locked position until a change is authorized. • The abbreviation LO or LC next to a valve on a flow diagram notes that a valve is to be locked open or locked closed.

(Continued)

Table A4.1 (Continued)

Type of equipment	Item	Notes
	Limit switches	• Limit switches are required when the position of a process or instrument valve must be known before a particular operational step can be taken or as part of an interlocked permissive sequence.
		• Information is provided by small limit switches on the valve housing. This is signified by using a Y as the final identification letter in an instrument bubble.
		• Only one switch is required to know whether a valve is fully open or fully closed. Two switches are needed to know if the valve is either fully open or fully closed.
	Automatic switch-over of pumps	• Automatic switch-over is required in critical services when a pumping operation must be maintained despite mechanical or electrical problems with the pump currently running.
		• Isolating valves about the spare pump are kept open and check valves provided in the discharge line from each pump.
		• A pressure-sensing device or flow switch in a common discharge line detects a pump failure, sounds an alarm, and automatically starts the reserve pump.
		• Pumps with severe but intermittent duties are frequently placed on alternate start-up whereby the pump started automatically is the one that had not previously been operating.
	Failure alarms	• Current to motors of critical pumps, compressors, agitators, or similar equipment is sometimes monitored to indicate the status of the process or of the equipment itself.
		• Engineering flow diagrams should reflect the presence of current indicators and accompanying low or high alarms.
	Interlocks	• Often the operation of a piece of equipment depends on another function in the process before it can begin, continue or terminate its operation.
		• The permissive conditions are covered by the appropriate instrumentation or notes on the engineering flow diagrams. Complex interlocks are indicated by an appropriate symbol to show that a separate logic diagram prepared by the instrument engineer covers the instrumentation and sequence of operations.

(Continued)

Table A4.1 (Continued)

Type of equipment	Item	Notes
Miscellaneous		Flow diagrams show many miscellaneous and specialty items to complete a project. Some of the more frequently encountered features are reviewed here:
	Expansion tanks	• Expansion tanks are required to relieve thermal expansion when liquid-filled sections of piping runs can be isolated and subsequently heated. • They may not be needed with water, as developed pressure is relieved by minor distortions of piping plus a small leakage at gaskets or by the use of a thermal relief valve. • They can be used to reduce water hammer or induced shock waves for any fluid. • Units are shown on a flow diagram with their dimensions or as a note giving a make and model number if they are to be purchased units.
	Backflow preventers & air gaps	• Backflow preventers are required by authorities supplying municipal water to prevent contamination of the water supply. • Backflow preventers or air gaps are used within a plant to ensure that contaminants do not enter the potable, locker-room or safety shower water systems. • A backflow preventer is a commercial item consisting of a series of special check valves and pressure-differential chambers to ensure the absence of backflow. An air gap is an arrangement whereby the line bringing incipient process or plant water from the municipal supply is terminated several inches above the tank that is to supply the process of plant water..
	Safety showers, eyewashes	• These are required whenever hazardous liquids or solids are being unloaded or handled. • They are usually purchased as a single combination unit. • Units are sized to give about $8m^3/h$ to the shower portion and $1.5m^3/h$ to the eyewash for a water pressure of 2 Bar. If the pressure is less, it must be boosted, and it if is too great, units must be supplied with the requisite restriction orifices.

(Continued)

Table A4.1 (Continued)

Type of equipment	Item	Notes
	Utility service stations	• The units should be connected directly to the potable water supply. If they are taken from a pumped process- or plant-water supply, their feed lines should be taken immediately after the discharge of the process-water pump. A backflow preventer isolates the shower–eyewash–water takeoff point from any process or plant users. • Units located outdoors must be protected from freezing. This is accomplished by having an underground water-supply assembly designed so that water remains below the freezing level until the unit is activated and residual water at the end of use passes into a prepared drain at the base. Aboveground supply piping is usually protected by electric self-limiting tracing or electric induction heating. The shower or eyewash unit itself is then supplied with tracing or induction heating. • Safety-shower and eyewash units are shown on engineering flow diagrams near the equipment with which they are needed with a reference to the utility flow diagram containing the distribution header. • Symbolic containers showing dilute acetic acid or carbonate solutions are often shown next to an eyewash to denote their presence for swabbing purposes. • It is good engineering practice to place utility hose stations at convenient locations throughout a plant. There should be air for such tasks as blowing dirt or water from equipment, unplugging or cleaning out lines, and operating pneumatically driven mechanical equipment or tools; water for housekeeping and emergency cooling; and steam to thaw or clean lines and equipment. Steam and water can be brought together to form a separate hot-water station. • Stations are sited on equipment-arrangement drawings and given identifying numbers. Each station is shown on the respective utility flow diagram and identified by the assigned number.

(Continued)

Table A4.1 (Continued)

Type of equipment	Item	Notes
	Fire protection	• Air, steam and water pressure for normal usage is limited to 2 Bar by pressure reducers to prevent a pressure hazard caused by high-velocity fluid impinging on the skin or by the whipping action of hoses. High-pressure air required to drive pneumatic motors or tools should have special connectors. • Hoses are typically 10 m long. If a series of hose stations is required, they can be located approximately 20 m apart. • Fire protection is required in operating and storage areas if flammable ingredients are used in the process or if gaseous or liquid fuels are present. • It is provided for control rooms, laboratories, maintenance shops, and office areas. • Although chemical canister units are provided, fire hydrants with fire-water pumps and underground fire-water mains are frequently included. The number and location of fire hydrants are set in accordance with local ordinances. • A method to reduce danger from a fire within a large oil storage tank is to sparge steam into the vapor space to displace air and snuff out the fire by smothering it.
	Seal fluid systems	• Reference is made at appropriate seal or packing locations on engineering flow diagrams to the seal or flushing fluid system required at the respective application point. • A notation lists the flow diagram which presents the distribution header and the line-identification codes of lines bringing and returning the fluid. • A separate section on a utility flow diagram should illustrate a schematic of the typical instrumentation and piping required at each application point for the various systems, list the various instruments and line numbers, and show the distribution systems with pertinent information in the line table.

APPENDIX 5

Teaching Practical Process Plant Design

INTRODUCTION

I cannot claim to have perfected design teaching though, after a few years of development, our students can now produce deliverables which look like their professional counterparts, and they understand how to work in teams to tight deadlines to produce them. They also have an idea of what it is like to be an engineer, as I make the context and frame of reference of the exercises as realistic as possible.

We teach process plant design from week 1 of year 1 in both the Chemical and Environmental Engineering courses at Nottingham, in line with the IChemE's wishes. The presently commonplace approach of waiting until year 3 to ask students to design something, having only received two years of noncontextualized natural science content, does not in my view work at all well. Many hours of guided and realistic practice are what are needed for good basic competence.

The idea that process plant design is founded in natural science is given the lie by the poor quality output of students who are set the challenge of a year 3 process plant design based only upon math, science, and a copy of Sinnot & Towler. This is not requiring them to carry out process plant design, this is requiring them to invent the discipline from scratch.

Since their lecturers are expert in research rather than design, the traditional approach produces a largely written document, with no recognizable engineering deliverables, a smattering of rote hand calculations straight from Sinnot & Towler, or worse still, a screen dump of a simulation program. Great emphasis may be given to proper referencing of the primary literature sources the design is based on, in a document which bears a strong resemblance to a research paper.

The students may well be asked to do the exercise in a group, but students and staff often collude to make it into a linked set of individual exercises. The smartest students will have designed the reactor, and the least able a few storage tanks or a heat exchanger. No one will have designed a complete plant. On questioning, it will become clear that none of the students understand anything about the process at a whole system level, which was supposed to have been the whole point of the exercise.

I have even seen process plant design courses which were assessed by formal examination, especially in the Far East. Process plant design ability cannot really be measured by traditional formal individual examination. It is too time-consuming, too collaborative, and too creative to be measured using a technique which is good for measuring an individual's ability to recall facts and carry out rote calculation under great pressure.

I prefer to assess a student's ability to undertake a design exercise, which will be collaborative in professional practice, by getting them to undertake collaborative design exercises based on real plants I have designed.

There is, however, a potential difficulty with this approach which needs to be managed—freeriders. Luckily for us, the UK Higher Education Academy's Engineering Centre financed the production of a dedicated software tool to help with this.

PEDAGOGY

I am if anything less keen on pedagogic philosophizing than I am on the philosophy of engineering, but there are a few useful ideas knocking about. These usually, however, need separating from the politicized and doctrinaire "theory" which they come freighted with (even some philosophers agree that philosophy is a waste of time—see Brennan in "Further Reading").

I offer a series of papers in "Further Reading" for those who want to see "research" backing of my teaching approach in peer-reviewed literature.

In summary, after investigation, I have concluded that pedagogic research is all too frequently worthless, the introduction of ideas from pedagogic theory is usually counterproductive, and there is not a single agreed fact in the entire field.

Educationalists may still be stuck at the stage of defining their terms and methodologies, but even a blind sow finds the odd acorn. Here are a few ideas which I have found worthwhile if used with caution. Note, however, that to the extent that they are useful they are also bleedin' obvious.

X-based learning/XBL (where X represents a word like "problem," "experience," or whatever) has come in and out of fashion since the nineteenth century, mainly because it sounds like a great idea, accords with some people's political leanings, and students like it, but it basically doesn't work.

It takes a lot of lecturer's time to do well, maybe five times as much as a traditional approach. It is therefore incredibly ineffective and inefficient as a teaching method, with one possible exception, its use for teaching professional skills. I use it only for this purpose.

"Constructive Alignment" suggests making as direct a link as possible between teaching and assessment techniques and the thing being taught. If I want to know how good a student is at designing plants, why not ask him to design a plant and assess the output?

The CDIO (Conceive, Design, Implement, Operate) movement recognizes many of the problems with engineering education which I do but, being pedagogues rather than engineers, they mix the principles of liberal arts education and doctrinaire philosophizing into their suggested solution to the problem. They have, however, rightly identified many of the problems, and anyone wanting a research-based backing of my suggestions could look to the website at http://www.cdio.org/.

"Good assessment practice"—I give an explicit explanation of learning objectives, both at the start of the course and along with each coursework making it clear what I am trying to have the students do. Each coursework also comes with a list of the relevant "assessment criteria" (which have to be met to obtain a passing grade) and "grading criteria", which describe how I will differentiate quality of work at grades above the bare pass.

My grading criteria are broadly aligned with scales like the Bloom taxonomy with one major difference. I do not reward the "extended abstract," which I consider an artifact of the discipline of those who devise such taxonomies. Piling supposition on top of philosophy on top of speculation is considered very clever indeed in the liberal arts and humanities. Such a fragile approach is not, however, well thought of in engineering. Rewarding it in my marks scheme would be counterproductive.

"Good feedback" is very important in teaching a skill like design. Things mentioned in the pedagogic literature like avoiding empty praise and empty criticism, and instead giving specific advice on how to do better next time, as well as explicit guidance on how I am going to determine grade boundaries, improves student work incrementally with each coursework. I do not allow my clear guidance on grade boundaries to be an opening for mark negotiation. I am teaching design, not negotiation.

"Formative assessment" sounds like a great idea to teach things like design which require a lot of practice. The only problem I have found with it is its complete impracticality. The vast majority of today's students seem to be what are called "strategic learners." They basically won't do a thing unless there are degree credits associated with it, and they want the amount of marks available to correlate with the hours of work pretty closely. I need to set marked assessments to get students to put the required hours of practice in.

One last thing—as the paper by Prince in "Further Reading" shows, the lecturer's own research needs to be kept out of design teaching. Contrary to popular belief, there is at best no correlation between research and teaching excellence. This is if anything more true if the subject of your research is something you call "process design."

METHODOLOGY

My teaching essentially supports a series of group courseworks of increasing difficulty. Working on these courseworks will take up the great majority of students' time. I don't do much traditional lecturing (and I'll do less still when this book comes out).

I have a teaching fellow with a lot of professional drafting experience who teaches the students how to use AutoCAD and to some extent Excel in computer lab sessions. Between the two of us there is a lot of contact time, around six hours per week, and I might also see groups who are having difficulty in my office for another half an hour a week.

I have classes of around 150, and I set three courseworks per module with maybe six deliverables each. Group work allows me to reduce the amount of marking I have to do to practical levels. Splitting the class into 20 groups reduces my marking load by a factor of seven.

I set the coursework with tight deadlines, and I do not negotiate on workload, group composition, deadlines, or marks. This allows me to get marks and feedback back to students in a day or two, so that they will not repeat their mistakes on the new coursework they will have received.

The UK's Higher Education Academy produced a very useful and robust tool for peer assessment called WebPA which I use to prevent free-riding in the group work exercises.

I have a number of collaborating process plant designers who help to deliver lectures, set assessments, and conduct design reviews with our students in years 1−3. Our year 3 design project incorporates two group design reviews supervised by highly experienced chartered engineers with decades of professional experience. The output from all of our design courses is professional front end engineering design (FEED) study documentation; and design reviews aim to produce professional quality documents (albeit early-career version).

There are a number of modules which have to be integrated with the process design module for best results, most notably process control, materials, fluid mechanics, and separation processes modules.

We have also stopped teaching students how to use modeling and simulation programs until year 3, and do not allow their use in year 3 projects. Early access to simulation programs has the same effect on process design ability as access to calculators has on mental arithmetic.

As I said in the last section, this is not a time-efficient approach. It takes up a lot of my time and a lot of the students' time. But, as I am making them into engineers, I think it is worthwhile for all of us.

EXERCISES

I have included in this section the exercises I use in my design teaching. All the courseworks are real plants I (or in one case a visiting professional) have actually designed.

I have not included model solutions, as I recommend this text to my students, and in any case there are no right answers. Even my own design solutions to these problems are not "right answers," as they were based on the state of the art at the time. I would do them differently (and better) today.

I have, however, included along with the courseworks my marking criteria (which I give to students along with the question when setting coursework) which explain to both students and lecturers what I am after from the exercises.

  1. Class exercise—The marshmallow challenge
  2. Class exercise—A 5 m^3 tank
  3. Coursework 1—Ortoire water treatment plant
  4. Coursework 2—Alnwick Castle water feature
  5. Coursework 3—Upgrading Jellyholm water treatment plant
  6. Class Exercise—Fun size creativity
  7. Coursework 4—Groundwater pilot plant
  8. Coursework 5—Supercritical water oxidation plant
  9. Class exercise—Pharmaceutical Intermediate Bulk Container (IBC)
  10. Coursework 6—Pharmaceutical aerosol manufacture

Class exercise: The marshmallow challenge

"The task is simple: in eighteen minutes, teams must build the tallest free-standing structure out of 20 sticks of spaghetti, one yard of tape, one yard of string, and one marshmallow. The marshmallow needs to be on top."

See http://marshmallowchallenge.com for details and supporting video.

A 40" tower is thought to be the world record, 20" is average.

Class exercise: A 5 m^3 tank

Very early in the course, students are asked to give the optimal height to diameter ratio for a tank for liquid with a 5 m^3 capacity made of 3 mm thick stainless steel.

Though they know all the necessary mathematics to solve the problem, they will struggle to make the necessary assumptions, and to frame the problem in a way which allows it to be solved using mathematics.

The instructor should not give them clues, or even tell them whether it has an open top or bottom unless asked.

There are very different solutions based on which assumptions are made, and some sets of assumptions, while themselves reasonable, yield no useful mathematical solutions.

Students who complete the exercise should be asked to comment on how many significant figures are meaningful in the answer if 3 mm stainless steel costs £2/m^2.

The problem is very useful for demonstrating the difference between engineering and math, and for introducing sensitivity analysis to an audience of mathematically competent engineering incompetents like freshers (or university lecturers).

Coursework 1: Ortoire water treatment plant
Task

1. Produce, as individuals, a single tab Excel spreadsheet which shows a rough initial design for a water treatment plant taking 15MLD of water from a lake at 750 mAOD, passing it through settlement tanks on to pressure sand filters, and from there into a chlorine contact tank to TR60 of 30 min HRT and 1 m water depth at 575 mAOD. Distance from lake to CCT is 1,500 m. Fall on ground is constant along this 1,500 m.
2. Numbers and diameters for circular settlement tanks and sand filters, external and internal dimensions of CCT, and theoretical and actual pipe diameters sizes will be needed to be calculated as a minimum.
3. Produce a single Excel file for your group which combines all of your individual submissions, with the submission you have as a group agreed is the best at the front.
4. Use your calculated sizes of units to produce a group scale drawing showing as much detail as you can in both plan and elevation.
5. Peer assess each other using WebPA.

Rules of thumb for design
- Surface loading settlement tanks 1 m/h
- HRT settlement tanks 2 h
- Max diameter settlement tanks 30 m
- Surface loading sand filters 15 m/h
- Sand filter depth 2 m
- Max diameter sand filters 3 m
- Static head required for sand filter operation 10 m
- Superficial velocity pipework 1.5 m/s

TLAs
- AOD—above ordnance datum
- HRT—hydraulic residence time
- CCT—chlorine contact tank
- MLD—megaliters per day

Learning outcomes

D1: Define a problem and identify constraints.

D2: Design solutions according to customer and user needs.

S1: Knowledge and understanding of commercial and economic context of chemical/environmental engineering processes.

P1: Understanding of and ability to use relevant materials, equipment, tools, processes, or products.

P5: Ability to use appropriate codes of practice and industry standards.

P8: Ability to work with technical uncertainty.

Assessment criteria

1. Apply British Standards
2. Size unit operations via Rules of Thumb
3. Undertake hydraulic calculations
4. Produce appropriate plant layout drawing (plan and elevation)

Grading criteria

ALL Assessment Criteria passed to a satisfactory level	**3RD**
Produce Excel spreadsheet which sizes unit operations and pipework with evidence of some analysis. Integrate into system design and produce a drawing to BS that clearly shows design intent.	**2:2**
Produce Excel spreadsheet which sizes unit operations and pipework with evidence of detailed analysis. Integrate into system design and produce a drawing to BS that clearly shows design intent. Cost, safety, and robustness are considered.	**2:1**
Produce Excel spreadsheet which sizes unit operations and pipework with evidence of critical analysis. Integrate into system design and produce a detailed drawing to BS that clearly shows design intent. Cost, safety, and robustness are analyzed.	**1ST**

Teaching Notes: You can make the exercise harder by specifying the quantity of water produced rather than the quantity input, requiring an iterative solution to the mass balance.

Coursework 2: Alnwick Castle water feature

You have been engaged to provide hydraulic and process design of a new water feature at Alnwick Castle by Company X, on behalf of the Duchess of Northumbria.

Your responsibility is to make sure that the water in the feature is suitably treated, and that the required effects are achieved.

The drawing provided shows the Grande Cascade, which is the centerpiece of the gardens (YouTube has plenty of videos of it in action). Hidden underneath the cascade are the plant rooms containing the pumps and treatment plant which make the feature work. Your job is to design the required plant and pumps, and fit them in into the plant rooms.

You need to submit your calculations as an Excel spreadsheet, a P&ID, and a modified version of the GA provided (note that all working water levels are to be marked on the GA), along with manufacturers datasheets for any equipment used.

The Company X Water Features Design Manual provided should facilitate your design, but additional marks are valuable for a more sophisticated approach. (I wrote the design guide to allow plumbers to design water features—I expect more from you.)

Learning outcomes

D1: Define a problem and identify constraints.
D2: Design solutions according to customer and user needs.
S1: Knowledge and understanding of commercial and economic context of chemical/ environmental engineering processes.
P1: Understanding of and ability to use relevant materials, equipment, tools, processes, or products.
P5: Ability to use appropriate codes of practice and industry standards.
P8: Ability to work with technical uncertainty.

Assessment criteria

1. Apply British Standards
2. Size unit operations via Rules of Thumb
3. Undertake hydraulic calculations
4. Produce appropriate plant layout drawing (plan and elevation)
5. Produce P&ID

Grading criteria

ALL Assessment Criteria passed to a satisfactory level	**3RD**
Produce Excel spreadsheet which sizes unit operations and pipework with evidence of some analysis. Integrate into system design and produce drawings to BS that clearly show design intent.	**2:2**
Produce Excel spreadsheet which sizes unit operations and pipework with evidence of detailed analysis. Integrate into system design and produce a drawings BS that clearly show design intent. Cost, safety, and robustness are considered.	**2:1**
Produce Excel spreadsheet which sizes unit operations and pipework with evidence of critical analysis. Integrate into system design and produce detailed drawings to BS that clearly show design intent. Cost, safety, and robustness are analyzed.	**1ST**

Teaching notes: Students will be able to find out quite a lot of data on this real project on the internet. A good solution will involve taking publicly available data on flows and appearance and flowrates calculated in a number of ways to size the pumps and treatment plant.

Coursework 3: Upgrading Jellyholm water treatment plant
Scenario: Jellyholm water treatment works

The Jellyholm water treatment plant in Sauchie (FK10 3AZ; Figure A5.1) is fed with water from the nearby Gartmorn reservoir (Grid Ref NS 92000 94200). The reservoir

Figure A5.1 Jellyholm water treatment works. *Copyright Image reproduced courtesy of Doosan Enpure Ltd.*

is subject to periodic algal blooms, and you have been commissioned to upgrade the plant to handle these.

The plant feeds housing which has lead piping, and orthophosphate dosing is to be provided to prevent lead dissolving into the drinking water on its way to supply.

The assignment is to carry out as detailed a design as you can of an upgrade to the treatment works. You can if you wish follow what was done as shown on the case study document, or you can offer an alternative approach.

Deliverables

1. P&ID
2. GA
3. Hydraulic Calculations, from dam to treated water tank
4. Process Design Calculations including Mass Balance
5. Control Philosophy
6. Approximately 2,000-word proposal for your plant upgrade including capital and running cost estimate, and justification of design choices
7. WebPA assessment of yourself and fellow group members
   For avoidance of doubt:

   All drawings are to be in AutoCAD format. All spreadsheets are to be in MS Excel. Report in MSWord. Your drawings should be to British Standards. They should be neat, clear, and accurate. Your report should also be neat, clear, concise, and accurate.

   If you don't understand any part of this assignment, ask me to explain. If you cannot see how to do some or all of the task set, ideally you would ask among your group first, then see if you can find the answer by research, then ask me.

Learning outcomes

D1: Define a problem and identify constraints.

D2: Design solutions according to customer and user needs.

S1: Knowledge and understanding of commercial and economic context of chemical/environmental engineering processes.

P1: Understanding of and ability to use relevant materials, equipment, tools, processes, or products.

P5: Ability to use appropriate codes of practice and industry standards.

P8: Ability to work with technical uncertainty.

Assessment Criteria

1. Apply British Standards
2. Size unit operations via Rules of Thumb
3. Undertake mass balance and hydraulic calculations
4. Produce appropriate plant layout drawing (plan and elevation)
5. Produce P&ID
6. Produce Proposal
7. Produce Control Philosophy

Grading criteria

ALL Assessment Criteria passed to a satisfactory level	**3RD**
Produce Excel spreadsheet which sizes unit operations and pipework with evidence of some analysis. Integrate into system design and produce drawings to BS that clearly show design intent. Produce Control Philosophy which explains control system design intent. Produce proposal document which explains all proposed refurbishments.	**2:2**
Produce Excel spreadsheet which sizes unit operations and pipework with evidence of detailed analysis. Integrate into system design and produce a drawings BS that clearly show design intent. Produce Control Philosophy which clearly explains control system design intent, and adequately considers most common system disturbances. Produce proposal document which clearly explains all proposed refurbishments. Cost, safety, and robustness are considered.	**2:1**
Produce Excel spreadsheet which sizes unit operations and pipework with evidence of critical analysis. Integrate into system design and produce detailed drawings to BS that clearly show design intent. Produce Control Philosophy which clearly explains control system design intent and adequately considers many system disturbances. Produce proposal document which clearly explains all proposed refurbishments. Cost, safety, and robustness are analyzed.	**1ST**

Coursework 4: Groundwater pilot plant

A temporary groundwater treatment plant has been running for some time on a site, and it is time to propose a permanent replacement. A pilot trial has been undertaken to see if alternative technologies to that used for the temporary plant are viable.

Using the temporary and pilot plant data provided, choose a suitable mixture of technologies, and produce a conceptual design with GA and P&ID appropriate to the site shown on the drawing provided.

It should be noted that the approach used on the real plant may not be the optimal one based on the data provided to you.

Deliverables

Excel Spreadsheet with as a minimum mass balance, unit operation sizing, and hydraulic calculations.

AutoCAD General Arrangement Drawing.

AutoCAD P&ID.

Learning outcomes

D1: Define a problem and identify constraints.

D2: Design solutions according to customer and user needs.

S1: Knowledge and understanding of commercial and economic context of chemical/ environmental engineering processes.

P1: Understanding of and ability to use relevant materials, equipment, tools, processes, or products.

P5: Ability to use appropriate codes of practice and industry standards.

P8: Ability to work with technical uncertainty.

Assessment criteria

1. Apply British Standards
2. Analyze data provided to choose technologies
3. Size unit operations via Rules of Thumb
4. Undertake hydraulic calculations
5. Produce appropriate plant layout drawing (plan and elevation)
6. Produce P&ID

Grading criteria

ALL Assessment Criteria passed to a satisfactory level	**3RD**
Produce Excel spreadsheet which analyses data, sizes unit operations and pipework with evidence of some analysis. Integrate into system design and produce drawings to BS that clearly show design intent.	**2:2**
Produce Excel spreadsheet which analyses data, sizes unit operations and pipework with evidence of detailed analysis. Integrate into system design and produce drawings to BS that clearly show design intent. Show how cost, safety, and robustness have been considered.	**2:1**
Produce Excel spreadsheet which analyses data, sizes unit operations and pipework with evidence of critical analysis. Integrate into system design and produce detailed drawings to BS that clearly show design intent. Show how cost, safety, and robustness have been analyzed.	**1ST**

Class exercise: Fun-size creativity

After a short lecture on creative systems including the McMaster 5-point strategy, students are invited to solve a problem which proved very difficult in practice—producing "fun size" Mars bars. The problem is that the Mars bar filling sticks to the blades used to cut it up into small pieces.

Coursework 5: Supercritical water oxidation plant

Task

Produce, as individuals, multitab Excel spreadsheets which show a rough initial design for a supercritical water oxidation (SCWO) nanoparticle production plant comprising as a minimum the following elements, as shown on the PFD supplied:

1 No. DI water stock tank, dosing pump, and valves delivering up to 3 m^3/h

1 No. hydrogen peroxide stock tank, dosing pump(s), and valves delivering up to 1 m^3/h

1 No. metal salt 1 stock tank, dosing pump(s), and valves delivering up to 1.5 m^3/h

1 No. metal salt 2 stock tank, dosing pump(s), and valves delivering up to 1 m^3/h

1 No. capping agent stock tank, dosing pump(s), and valves delivering up to 1 m^3/h

1 No. electrical process heater

1 No. preheat heat exchanger

1 No. reactor

1 No. energy recovery heat exchanger

1 No. postfilter

2 No. cooling water pumps each delivering 6.5 m^3/h

Cooling tower

Cooling water filter

1. Sizes of these elements and actual pipe diameters will be needed to be calculated as a minimum.
2. Produce a single Excel file for your group which combines all of your individual submissions, with the submission you have as a group agreed is the best at the front.
3. Use your calculated sizes of units to produce individual scale drawings (GA) showing as much detail as you can in both plan and elevation, choose one to represent the group, and submit all as a single file with the chosen submission clearly marked.
4. Produce an individual P&ID showing as much detail as you can, choose one to represent the group, and submit all as a single file with the chosen submission clearly marked.
5. Peer assess each other using WebPA.

Learning outcomes

D1: Define a problem and identify constraints.

D2: Design solutions according to customer and user needs.

S1: Knowledge and understanding of commercial and economic context of chemical/ environmental engineering processes.

P1: Understanding of and ability to use relevant materials, equipment, tools, processes, or products.

P5: Ability to use appropriate codes of practice and industry standards.

P8: Ability to work with technical uncertainty.

Assessment criteria

1. Apply British Standards
2. Size unit operations via Rules of Thumb
3. Undertake mass and energy balance and hydraulic calculations
4. Produce appropriate plant layout drawing (plan and elevation)
5. Produce P&ID

Grading criteria

ALL Assessment Criteria passed to a satisfactory level	3RD
Produce Excel spreadsheet which sizes unit operations and pipework with evidence of some analysis. Integrate into system design and produce drawings to BS that clearly show design intent.	2:2
Produce Excel spreadsheet which sizes unit operations and pipework with evidence of detailed analysis. Integrate into system design and produce a drawings BS that clearly show design intent. Cost, safety, and robustness are considered.	2:1
Produce Excel spreadsheet which sizes unit operations and pipework with evidence of critical analysis. Integrate into system design and produce detailed drawings to BS that clearly show design intent. Cost, safety, and robustness are analyzed.	1ST

Class exercise: Pharmaceutical intermediate bulk container (IBC)

You are considering the design of a new dispensary for a solid dose facility. The following formulation is to be dispensed into IBCs within the dispensary.

Material	Quantity to be added (kg)	Bulk density (kg/l)	Occupation exposure limit
Lactose mono SDS	45.0	0.625	10 mg/m^3
Lactose	220.0	0.625	10 mg/m^3
Hydroxy cellulose	15.0	0.500	10 mg/m^3
Carboxy methyl cellulose	23.0	0.500	10 mg/m^3
Micronized active ingredient	30.0	0.150	5 μg/m^3

Figure A5.2 gives the typical dimension of the IBC that the client wishes to use. The dimensions shown are fixed and cannot be changed. This configuration is based on filling the IBC so that the top of the repose-cone of powder does not go any higher than the top of the straight section of the IBC. Calculate the length of straight side shown and hence the overall height of the IBC. Show your working and assumptions.

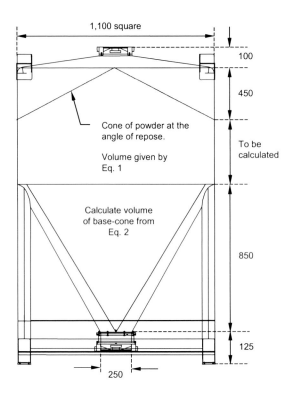

Figure A5.2 IBC.

Teaching Notes: This is like a more sophisticated version of the 5 m^3 tank exercise, which I give to second year and MSc students as a preparation for coursework 6. Even if you give them the formulae and simplifying assumptions below, they tend to make quite heavy weather of it.

It can be interesting to use figures rounded to 2—3 significant figures at the start of the exercise, to illustrate the problems of premature rounding.

The volume of repose-cone can be calculated from:

$$\text{Volume} = 1/3(\text{Base Area} \times \text{Height}) \tag{A5.1}$$

The volume of base-cone of the IBC can be calculated from:

$$\text{Volume} = 1/3(A_1 + A_2 + \sqrt{(A_1 \times A_2)} \times \text{Height}) \tag{A5.2}$$

where $A_1 =$ the cross-sectional area of the top of the base-cone and $A_2 =$ the cross-sectional area of the bottom of the base-cone. Assume that the bottom of the base-cone has a square cross section and ignore the transition to the circular outlet. Also assume that this square has the same length and breadth as the outlet valve diameter.

Coursework 6: Pharmaceutical aerosol manufacture
General requirements

No.	Description	Acceptance criteria
1.	Product contact surfaces—metallic materials of construction.	Stainless steel type EN 1.4435
2.	Product contact surfaces—polymeric materials	PTFE or PVDF compliant with ASME BPE 2012 Part PM
3.	Product contact surfaces—elastomeric materials	Perfluorelastomer, silicone, or PTFE encapsulated EPDM compliant with ASME BPE 2012 Part PM
4.	Product contact surfaces—metallic materials surface condition.	Compliant with ASME BPE 2012 Table SF-2.2-1
5.	Product contact surfaces—metallic materials surface finish.	Electro-polished to 0.38 μm. Compliant with ASME BPE Table SF-2.4-1
6.	Product contact surfaces—nonmetallic materials surface condition.	Compliant with ASME BPE 2012 Table SF-3.3-1
7.	Material certificate of compliance.	EN 10204 Type 3.1
8.	Product contact piping.	OD hygienic tubing—True imperial dimensions
9.	Product piping couplings	Hygienic union couplings compliant with EN 11864
10.	Product contact liquids and clean steam—valves	Weir-type diaphragm compliant with ASME BPE SG-2.3.1.2

Note: Product contact surfaces include all clean utilities downstream of the final sterilizing grade filter, propellant downstream of the final sterilizing grade filter, and clean in place systems.

Dispensary operations

No.	Description	Acceptance criteria
1.	Active ingredient maximum water content	2% w/w
2.	Active ingredient maximum ethanol content	1% w/w
3.	Active ingredient drying time.	15 h maximum
4.	Active ingredient dryer end point	80°C at 20 mbar abs
5.	Active ingredient dryer yield	>98% mass basis
6.	Active ingredient median particle size	2 μm mass basis
7.	Active ingredient milling yield	>98% mass basis
8.	Active ingredient CIP fluid	Purified water BP
9.	Active ingredient CIP fluid flow	35 l/min/m of vessel circumference with a static spray device
10.	Active ingredient CIP final rinse liquid.	Purified water BP
11.	Active ingredient dispensing accuracy	± 0.1% mass basis
12.	Excipient dispensing accuracy	± 0.5% mass basis
13.	Lubricant maximum water content	1% w/w
14.	Sweetener maximum water content	0.5% w/w
	Flavor maximum water content	0.5% w/w

Propellant storage and distribution

No.	Description	Acceptance criteria
1.	Design pressure—maximum	Vapor pressure at maximum design temperature plus 1 bar g
2.	Design pressure—minimum	Full vacuum
3.	Design temperature—maximum	50°C
4.	Design temperature—minimum	−20°C
5.	Tank capacity	Tanker load plus 5,000 l
6.	Temperature of propellant transferred to manufacturing	20 ± 1°C
7.	Transfer rate to manufacturing	To take no more than 20 min
8.	Level indication accuracy	± 1%
9.	Pressure relief requirements	Duplex pressure relief valve
10.	Pump head	Static head 10 m Piping equivalent length 300 m Heat exchanger pressure drop max 0.5 bar

11.	Pump NPSH	To avoid cavitation at all times
12.	Piping upstream of sterilizing filter—metallic material	Stainless steel type EN 1.4404
13.	Pump mechanical seal	Sealess magnetic seal
14.	Piping upstream of sterilizing filter—nonmetallic material	PTFE or EPDM
15.	Piping upstream of sterilizing filter—piping standard	Flanged to BS 4504
16.	Piping upstream of sterilizing filter—valves	Flanged Ball Valves
17.	Propellant filtration	Prefilter 5 μmFinal filter 0.2 μm sterilizing grade filter

Formulation

No.	Description	Acceptance criteria
1.	Design pressure maximum	Vapor pressure at maximum design temperature plus 1 bar g
2.	Design pressure minimum	Full vacuum
3.	Design temperature maximum	100°C
4.	Design temperature minimum	−20°C
5.	Maximum batch size	1,000 l
6.	Minimum batch size	500 l
7.	Vessel nominal volume	To the top of the heat transfer surface
8.	Vessel minimum stirred volume	50 l maximum. Batch shall remain homogeneous at minimum stirred volume
9.	Formulation temperature	20 ± 1°C
10.	Dispersion time	10 min at 400 rpm
11.	Mixing requirement	To achieve full suspension To achieve efficient heat transfer
12.	Heating requirement	Hold at 20°C during formulation and can filling Hold at 80°C during CIP and drying
13.	Cooling requirement	Cool to 20°C after drying in 30 minutes
14.	In-place cleaning fluid	2% w/w NaOH in purified water BP
15.	In-place cleaning final rinse	Purified water BP
16.	In-place cleaning temperature	80 ± 1°C
17.	In-place cleaning fluid flow	35 l/min/m of vessel circumference with a static spray device
18.	Heat transfer method	External dimpled jacket covering shell and bottom head
19.	Maximum pressure drop through heat transfer jacket	0.5 bar g
20.	Maximum variation of propellant concentration during can filling	± 1% of propellant concentration

21.	Maximum water concentration in final formulation	0.5% w/w
22.	Maximum oxygen concentration in final formulation	0.1% w/w
23.	Maximum nitrogen concentration in final formulation	0.1% w/w
24.	Vessel drainability	Fully drainable with no pools greater than 5 mm diameter after draining
25.	Mixing dead spots	No mixing dead spots, vessel surface shall be flush below the liquid level including the bottom outlet

Can filling

No.	Description	Acceptance criteria
1.	Can purge	2 can volumes minimum
2.	Filling temperature	20 ± 1°C
3.	Filling rate	30 cans per minute
4.	Rejects	2% maximum
5.	Samples	1 per 1,000 cans
6.	Overspray after filling head removed	1 off can valve capacity
7.	Function testing	3 valve actuations
8.	Maximum air concentration in suspension in the can	1% w/w
19.	Maximum water content in the suspension in the can	1% w/w
10.	Operating shift pattern	3 shifts per day Start Monday 6 pm Finish Friday 10 pm
11.	Formulation maximum hold time	40 h
12.	Air extraction rate each can filling booth	1,000 Nm^3/h
13.	Air extraction rate each function tester	1,000 Nm^3/h

Deliverables
- Full mass balance and energy balance
- Preparation of detailed PFD covering the whole process
- Detailed P&ID to industry standards covering equipment in group in No. 3 above
- Equipment sizing for all components
- GA

Equipment sizing should include
- Operational size, for example, liters, liters/min, kW, etc.
- Design capacity as above but with design margin added
- Approximate physical dimensions
- Materials of construction

- Design pressure
- Motor power
- NPSH for pumps, etc.

Complete industry standard data sheet for the formulation vessel, homogenizer, and agitator.

Blank data sheet to be provided.

Learning outcomes

D1: Define a problem and identify constraints.
D2: Design solutions according to customer and user needs.
S1: Knowledge and understanding of commercial and economic context of chemical/environmental engineering processes.
P1: Understanding of and ability to use relevant materials, equipment, tools, processes, or products.
P5: Ability to use appropriate codes of practice and industry standards.
P8: Ability to work with technical uncertainty.

Assessment criteria

1. Apply British Standards
2. Size unit operations via Rules of Thumb
3. Undertake mass and energy balance and hydraulic calculations
4. Produce appropriate plant layout drawing (plan and elevation)
5. Produce P&ID

Grading criteria

ALL Assessment Criteria passed to a satisfactory level	3RD
Produce Excel spreadsheet which sizes unit operations and pipework with evidence of some analysis. Integrate into system design and produce drawings to BS that clearly show design intent.	2:2
Produce Excel spreadsheet which sizes unit operations and pipework with evidence of detailed analysis. Integrate into system design and produce a drawings BS that clearly show design intent. Cost, safety, and robustness are considered.	2:1
Produce Excel spreadsheet which sizes unit operations and pipework with evidence of critical analysis. Integrate into system design and produce detailed drawings to BS that clearly show design intent. Cost, safety, and robustness are analyzed.	1ST

FURTHER READING

Biggs, J.B., 2003. Teaching for Quality Learning at University, second ed. Open University Press/Society for Research into Higher Education, Buckingham.

Brennan, J., 2010. Scepticism about philosophy. Ratio 23 (1), 1–16.

Hattie, J., 2009. Visible Learning; A Synthesis of over 800 Meta-analyses Relating to Achievement. Routledge, London.

Loddington, S., Pond, K., Wilkinson, N., Wilmot, P., 2009. A case study of the development of WebPA: an online peer-moderated marking tool. Br. J. Educ. Technol. 40 (2), 329–341.

Marzano, R.J., 1998. A Theory-Based Meta-analysis of Research on Instruction. Mid-continent Research for Education and Learning, Aurora, CO.

Nicol, D.J., Macfarlane-Dick, D., 2006. Formative assessment and self-regulated learning: a model and seven principles of good feedback practice. Stud. High. Educ. 31 (2), 199–218.

Orsmond P. (2004) Self- and peer-assessment: guidance on practice in the biosciences. Higher Education Academy Centre for Bioscience: Teaching Bioscience: enhancing learning series.

Prince, M.J., Felder, R.M., Brent, R., 2007. Does faculty research improve undergraduate teaching? An analysis of existing and potential synergies. J. Eng. Educ. 96 (4), 283–294.

Strivens, J., 2007. What theory should we use—if any? Interpreting "scholarship" on programs for new university teachers. PRIME 2 (2), 81.

The CDIO Initiative. Available from: <http://www.cdio.org/>.

Victoria University of Wellington, 2004. Group Work & Group Assessment. University Teaching Development Centre (Online). Available from: <http://www.utdc.vuw.ac.nz/resources/guidelines/GroupWork.pdf>.

GLOSSARY

There are a number of words and phrases which are not in common use and, more problematically, there are a number which are in common use, but mean different things to different people. In this glossary, I define the sense in which I am using them in this book. This is intended to be the most commonly held meaning.

Basic Engineering Design Data (BEDD) Information compiled to allow conceptual design in petrochemical sector

Best Practice The consensus heuristics of practitioners

Block Flow Diagram (BFD) Academic approximation of a PFD

BS British Standard

CAD Computer-aided design/drawing

Capex Capital expenditure

Conceptual Design The initial stage of design; content varies between sectors

Consultancy A company which offers advice, and rarely progresses design beyond conceptual stage

Contracting Company A company which contracts to build plants, and usually does its own detailed design

Control and Instrumentation Engineer A hybrid chemical/software engineer or sometimes instrument technician found mainly in petrochemical industry operating companies

CoV Co-efficient of variance

DCS Distributed Control System

DEFRA UK Department for Environment, Food & Rural Affairs

Deliverables Things delivered under a contract; in a plant design context, mainly drawings

Design Basis Information compiled to allow design at any stage, in general terms. See BEDD for a sector-specific exception

Design Philosophy Accounts of decisions on how a number of common design problems and issues will be handled during a design; best generated early in the design process

Designer Someone who designs a plant and, in the context of this book, is willing to be legally responsible for it

Dimensioned Drawing Drawing marked with dimensions of real-world counterparts of items illustrated. Not guaranteed to be a scale drawing

Draffie Draughtsman/woman

Due Diligence Generating sufficient certainty in your opinions, considering the potential downside if you are wrong

dxf Drawing Exchange Format, a file format developed by Autodesk (authors of AutoCAD) which allows usually less than perfect file sharing with other CAD programs

EA UK Environment Agency

EPC Engineering, Procurement and Construction company, aka "Contracting Company"

Engineering The profession of imagining and bringing into being a completely new artifact which safely, cost—effectively, and robustly achieves a specified aim

Engineering Science The application of scientific principles to the study of engineering artifacts

FEANI "Fédération Internationale d'Associations Nationales d'Ingénieurs"

FEED Front End Engineering Design—an initial design exercise

Functional Design Specification (FDS) A description in carefully chosen words of what a plant designer wants the software to do

General Arrangement (GA) Drawing A scale drawing which shows the layout in space of a plant; aka a "plot plan" among other things

HAZASS Hazard Assessment: Mecklenburgh's hazard evaluation/identification technique

HAZOP Hazard and Operability Study

HMI Human—machine interface

HSE Health Safety Environment or UK Health and Safety Executive

IPPC Integrated Pollution Prevention and Control

ISO International Standards Organization

Mogden Formula A formula used in the United Kingdom to calculate trade effluent charges

Natural Science The activity of trying to understand natural phenomena (cf. Engineering Science)

Olfactorithmetic The ability to detect an implausible numerical answer by "smell"

Operating Company A company whose main activity is managing and operating process plants

Opex Operating expenditure

Optimization Improving a process by balancing a number of variables against cost, safety, and robustness

Partial Design An academic approximation of parts of the design process which falls far short of total design (qv)

PERT Program/project evaluation and review technique. A program evaluation tool, allied with critical path analysis

Pinch Analysis A largely academic exercise to minimize resource usage

Piping and Instrumentation Diagram (P&ID) The process engineer's signature drawing, showing physical and logical interrelationships between process plant components

Piping Engineer Specialist in piping and sometimes plant layout used in some industries and countries to produce plant layouts

PLC Programmable Logic Controller—Industrial Computer

Plot Plan See GA

Precision To do with whether the instrument will give me the same reading against the same true value the next time I test it (though I do use it in other senses in the book)

Problem A design problem will require the use of engineering judgment and imagination to solve, as data and/or design methodologies are lacking

Process Design An abstract conceptual "design" of a chemical process with no real consideration of cost, safety, or robustness

Process Flow Diagram (PFD) A drawing which represents the mass balance, resembling in many ways a simplified P + ID

Process Intensification A largely academic conception of combining unit operations, or Trevor Kletz's term for what is now usually called minimization in inherent safety techniques

QA Quality Assurance

Repeatability Precision under tightly controlled conditions over a short time period

Reproducibility Variability over time

RTFM (In polite terms) Kindly Read the Manual

SCADA System control and data acquisition

Scale Drawing Drawing whose dimensions are consistently some ratio of the size of their corresponding real-world counterparts

State Of The Art The set of heuristics of a designer or designers

Task A design task involves using a well-established methodology and robust data to grind out an obvious answer

Tiffie Instrument technician

T'Internet The World Wide Web in the vernacular of the North of England

TLA Three-letter acronym—engineering joke

Total Design "Total Design is the systematic activity necessary, from the identification of the market/user need, to the selling of the successful product to satisfy that need—an activity that encompasses product, process, people and organization."—Stuart Pugh

Validation Ensuring software matches reality

Verification Ensuring software is free of coding errors

WOD Write-only documentation

INDEX

Note: Page numbers followed by "*f*" and "*t*" refer to figures and tables, respectively.

CPI Antony Rowe
Eastbourne, UK
April 29, 2015